Generative AI in Education

A Guide for Parents and Teachers

Paolo Narciso

Apress®

Generative AI in Education: A Guide for Parents and Teachers

Paolo Narciso
Seabrook Island, SC, USA

ISBN-13 (pbk): 979-8-8688-0843-2   ISBN-13 (electronic): 979-8-8688-0844-9
https://doi.org/10.1007/979-8-8688-0844-9

Managing Director, Apress Media LLC: Welmoed Spahr
Acquisitions Editor: Shivangi Ramachandran
Development Editor: James Markham
Project Manager: Jessica Vakili

Cover image by eStudioCalamar

Distributed to the book trade worldwide by Apress Media, LLC, 1 New York Plaza, New York, NY 10004, U.S.A. Phone 1-800-SPRINGER, fax (201) 348-4505, e-mail orders-ny@springer-sbm.com, or visit www.springeronline.com. Apress Media, LLC is a California LLC and the sole member (owner) is Springer Science + Business Media Finance Inc (SSBM Finance Inc). SSBM Finance Inc is a **Delaware** corporation.

For information on translations, please e-mail booktranslations@springernature.com; for reprint, paperback, or audio rights, please e-mail bookpermissions@springernature.com.

Apress titles may be purchased in bulk for academic, corporate, or promotional use. eBook versions and licenses are also available for most titles. For more information, reference our Print and eBook Bulk Sales web page at http://www.apress.com/bulk-sales.

Any source code or other supplementary material referenced by the author in this book is available to readers on GitHub (https://github.com/Apress). For more detailed information, please visit https://www.apress.com/gp/services/source-code.

If disposing of this product, please recycle the paper

Table of Contents

About the Author

 **Paolo Narciso** is the founder of Core Immersive Academy, a Maryland Institute College of Art finalist for Creative Entrepreneurship in 2022. He was the former head of Product and Program Development at AARP Foundation, a Washington, DC-based national non-profit. He holds a Doctorate in Education from Creighton University and sits on multiple technology boards, advising on the use of artificial intelligence, blockchain in healthcare, and how to secure property rights for marginalized populations.

About the Technical Reviewer

 **Dr. James McCoy** is a builder and innovator in emerging technologies, such as AI/ML, blockchain, and Web3, with over eight years of IT experience. He is currently an Emerging Technology Fellow at the US Census Bureau, where he applies his expertise in privacy enhancing technology and explainable AI to support the agency's data privacy and security initiatives.

James is also the founder and platform architect of Kutline, a web development company that creates cutting-edge web products and solutions for various clients and industries. As the technical owner and platform solutions architect, he leads the planning, designing, and integration of platform technologies and mentors and trains employees on system design, secure implementation, and AWS services. James is passionate about leveraging technology to solve complex problems, improve lives, and advance social justice.

Introduction: Embracing the Generative AI Revolution in Education

The classroom was silent except for the scratching of pencils and the occasional rustle of paper. A young student, eyes fixed on his notebook, diligently worked through a series of trigonometry problems. As the sunlight streamed through the windows, beads of sweat formed on his brow – not just from the warmth, but from the pressure of knowing tomorrow's quiz loomed. Then, as class ended, a wave of relief washed over him: calculators would be permitted for the test! That student was me, and little did I know how profoundly technology, beyond the calculator, would continue to shape education in the decades to come.

Fast forward to the present day. As I watch my granddaughters navigate their educational journeys, I find myself in a familiar yet vastly different landscape. The calculator, once a controversial tool in the classroom, now seems quaint compared to the array of digital devices at students' fingertips. Laptops, tablets, and smartphones have become ubiquitous, and a new technological frontier is emerging: Artificial Intelligence (AI) and its subset, Generative AI.

The parallels between the calculator's introduction and the current AI revolution in education are striking. Just as calculators once sparked debates about the potential erosion of basic math skills, AI now raises questions about critical thinking, creativity, and the very nature of learning itself. Yet, as history has shown, technological advancements, when properly integrated, can enhance rather than hinder education.

My own journey with technology in education began with those early days of calculator use. I remember the initial resistance from teachers, the strict rules against their use in certain classes, and the gradual acceptance as their benefits became apparent. What started as a tool for basic arithmetic evolved into sophisticated devices capable of complex computations. Suddenly, I found myself grappling with functions like Sine and Cosine – a challenge that pushed the boundaries of my calculator skills at that time.

Now, as I seek to support my granddaughters in their learning, I face a new challenge: understanding and leveraging AI in education. The leap from calculators to AI is monumental, yet the underlying principle remains the same – technology, when used thoughtfully, can be a powerful ally in the learning process.

This realization led me to dive deep into the world of Generative AI, exploring its potential to create personalized content, provide instant feedback, and adapt to individual learning styles. The more I learned, the more I saw the immense potential for AI to transform education, much like the calculator did in its time, but on a far grander scale.

However, I also recognized a gap. While there was an abundance of technical information about AI in education, there was a lack of accessible, engaging resources for families like mine who were trying to navigate this new terrain. This need sparked the creation of this book – a guide designed to help parents, grandparents, and educators understand and harness the power of AI in learning.

The Importance of Collaboration in AI-Assisted Learning

Before we delve into the specifics of AI in education, it's crucial to highlight the importance of collaboration between students, parents/guardians, and teachers in this new learning landscape. This book, *Generative AI in*

Education: A Guide for Parents and Teachers, is not just a book to be read passively; it's an interactive guide that encourages active participation from all stakeholders in a student's education.

Throughout this book, you'll find exercises and activities designed for students, parents/guardians, and teachers to engage with together. This collaborative approach is fundamental to the successful integration of AI in education for several reasons:

1. Shared Understanding: By working together, all parties can develop a common understanding of AI tools, their potential benefits, and their limitations. This shared knowledge base facilitates more effective communication and decision-making about AI use in learning.

2. Diverse Perspectives: Students, parents, and teachers each bring unique perspectives to the table. Students offer insight into their learning needs and preferences, parents provide context about home learning environments, and teachers contribute pedagogical expertise. This diversity of viewpoints enriches the AI-assisted learning experience.

3. Ethical Considerations: The ethical use of AI in education is a complex topic that benefits from multi-stakeholder discussion. By engaging in exercises together, students, parents, and teachers can explore ethical dilemmas and develop guidelines for responsible AI use.

4. Support and Guidance: As students navigate the world of AI-assisted learning, having support from both parents and teachers is invaluable. Collaborative exercises allow adults to guide students in developing critical thinking skills and responsible tech habits.

5. Bridging Home and School: AI-assisted learning often transcends traditional boundaries between home and school. By working together, parents and teachers can ensure continuity and consistency in how AI tools are used across different learning environments.

6. Adapting to Individual Needs: Every student is unique, and AI tools offer the potential for highly personalized learning experiences. Collaborative exercises help identify each student's specific needs and preferences, allowing for more tailored AI integration strategies.

7. Continuous Improvement: The field of AI in education is rapidly evolving. By engaging in ongoing collaborative activities, students, parents, and teachers can stay updated on new developments and continually refine their approach to AI-assisted learning.

As you progress through this book, we encourage you to approach the exercises and activities as opportunities for meaningful collaboration. Whether you're a student eager to explore new learning tools, a parent looking to support your child's education, or a teacher seeking to integrate AI into your curriculum, your active participation will enrich the learning experience for everyone involved.

A Journey Through the Book

Now, let's embark on our journey through the exciting world of AI in education. Here's what you can expect in the chapters ahead:

Chapter 1: Getting to Know Your AI Education Partner

In this chapter, we'll explore the current landscape of AI in education, discussing both the potential benefits and challenges of using AI as a study aid. We'll set the stage for what students and parents can expect from this book, providing a solid foundation for understanding the role of AI in learning.

Chapter 2: Staying Safe and Smart with AI

As we venture into the world of AI-assisted learning, it's crucial to address the ethical considerations that come with this powerful technology. This chapter emphasizes the importance of parental involvement in guiding the ethical use of AI. We'll provide age-appropriate guidelines for students in different grade ranges (3–5, 6–8, 9–12) and explore specific examples of ethical dilemmas and how to approach them. You'll also find sample prompts and exercises for discussing AI ethics with your child.

Chapter 3: The Magic of Prompts: Unleashing Generative AI's Potential

Here, we dive into the heart of interacting with AI tools – crafting effective prompts. We'll cover the basics of prompts, the art and science of creating them, and provide sample prompts and exercises for different subjects and grade levels. We'll also discuss the iterative process of refining prompts, address challenges and limitations of AI-assisted learning, and highlight the importance of collaboration and communication. The chapter concludes with a fun parent–student activity and a look ahead to the exciting world of prompt engineering.

Chapter 4: Your AI Toolkit: ChatGPT, Claude, and Google's Gemini AI

This comprehensive chapter introduces you to some of the most powerful generative AI tools available: ChatGPT, Claude, and Google's Gemini AI. We'll provide detailed examples and exercises for using each tool across different subjects and grade levels. You'll learn how to create effective prompts, understand the unique features of each AI assistant, and explore best practices for utilizing these tools in education. We'll also discuss potential limitations and ethical considerations to keep in mind.

Chapter 5: Bringing AI Home: Enhancing Learning Outside the Classroom

AI-assisted learning doesn't stop at the school gates. This chapter focuses on how to set up an AI-friendly learning environment at home. We'll explore strategies for using AI to support different learning preferences (visual, auditory, kinesthetic), discuss how to balance AI-assisted learning with traditional methods, and guide you in crafting a personalized AI-assisted learning plan.

Chapter 6: Teaming Up with Teachers: AI in the Classroom

Effective integration of AI in education requires collaboration between home and school. This chapter provides guidance on communicating with teachers about AI use in your child's learning process, aligning AI-assisted learning with classroom expectations and assignments, and advocating for the responsible use of AI in schools.

Chapter 7: Tracking Your AI Adventure: Progress and Impact

As with any educational tool, it's important to monitor progress and assess impact. In this chapter, we'll discuss methods for tracking student improvement with AI assistance, evaluating the effectiveness of AI-generated prompts and feedback, and addressing potential challenges. We'll also provide sample prompts and exercises for monitoring progress and assessing AI's impact on learning.

Chapter 8: The Future of Learning with AI

The field of AI in education is rapidly evolving. This chapter looks ahead to emerging trends and technologies, discussing potential long-term benefits and challenges. We'll explore how to empower students to thrive in an AI-driven world and provide sample prompts and exercises for discussing the future of AI with your child.

Chapter 9: Wrapping Up Your AI Journey

In our final chapter, we'll recap key takeaways from each chapter of the book and provide a roadmap for implementing the strategies we've discussed. We'll conclude with a call to action, encouraging readers to embrace AI as a powerful tool for learning and growth.

Appendix

Because the landscape of Generative AI is changing quickly, we've created an appendix to cover topics that frankly could be books on their own. Topics like how to use Generative AI for creating images, videos, pictures, graphs and charts will be covered in this appendix. Tools like Dall-E and Mindjourney will also be covered. Additionally, we've created a list of prompts that may be useful in kickstarting your journey in using Generative AI to create art and graphics.

Glossary

To ensure that all readers can fully engage with the content, we've included a comprehensive glossary explaining AI terms and concepts for students and parents.

Navigating the AI Revolution Together

As we begin on this journey through this book, it's important to remember that we're all learners in this new AI-enhanced educational landscape. Whether you're a student excited about new learning tools, a parent trying to support your child's education, or an educator looking to integrate AI into your teaching, this book is designed to meet you where you are and guide you forward.

The exercises and activities throughout this book are not just for show – they're carefully crafted opportunities for hands-on learning and exploration. We strongly encourage you to engage with these exercises, ideally in collaboration with others. Students, work through the activities with your parents or teachers. Parents, use these exercises as opportunities to connect with your child's learning journey. Teachers, consider how you might adapt these activities for your classroom or use them to spark discussions with parents about AI integration.

Remember, the goal isn't to become an AI expert overnight. It's to develop a working understanding of AI tools, their potential benefits, and their limitations in educational contexts. It's about learning to ask the right questions, think critically about AI-generated content, and use AI responsibly to enhance learning rather than replace critical thinking skills.

As you progress through the book, you might find some concepts challenging or encounter ethical dilemmas that don't have clear-cut answers. That's okay – in fact, it's expected. The field of AI in education is complex and rapidly evolving. What's important is that you're engaging with these ideas, discussing them with others, and developing your own informed perspectives.

Embrace the iterative nature of learning about and with AI. Just as we refine our prompts to get better responses from AI tools, we can refine our understanding and use of AI in education over time. Be prepared to experiment, make mistakes, learn from them, and try again.

Also, keep in mind that while AI tools are powerful, they're not a panacea for all educational challenges. They're most effective when used in conjunction with human guidance, critical thinking, and hands-on experiences. Throughout this book, we'll emphasize the importance of maintaining a balance between AI-assisted learning and other educational approaches.

A Personal Note

As I reflect on my own journey from struggling with a calculator in trigonometry class to exploring the frontiers of AI in education, I'm filled with a sense of wonder at how far we've come. Yet, I'm also reminded that the fundamental goals of education remain the same: to foster curiosity, critical thinking, and a love of learning.

My hope is that this book will serve as a bridge between the familiar world of traditional education and the exciting, sometimes daunting world of AI-enhanced learning. I hope it will empower you – whether you're a student, parent, or teacher – to approach AI with confidence, creativity, and a critical eye.

Remember, just as the calculator didn't replace the need for mathematical understanding, AI won't replace the need for human intelligence, creativity, and interpersonal skills. Instead, when used wisely, it can augment our abilities, free up time for deeper learning, and open up new avenues for exploration and discovery.

As you read this book, I encourage you to maintain an open mind, but also a questioning one. Engage with the ideas presented, try out the exercises, and most importantly, discuss what you're learning with others. Education has always been a collaborative endeavor, and in the age of AI, this collaboration becomes even more crucial.

So, are you ready to meet your AI partner in education? Are you prepared to explore how AI can transform the way we learn, teach, and think about education? If so, turn the page, and let's begin our adventure into the world of AI-assisted learning. Remember, in this journey, we're all learners, we're all explorers, and together, we can shape the future of education in the AI age.

Let's embrace this exciting new chapter in education, learning and growing together – one AI-assisted step at a time. Welcome to *Generative AI in Education: A Guide for Parents and Teachers*. Your adventure in AI-enhanced learning starts now!

CHAPTER 1

Getting to Know Your AI Education Partner

The bell rang, signaling the end of another school day. As students filed out of the classroom, 12-year-old Sophia lingered behind, her face etched with frustration. Her teacher, Mr. Johnson, noticed her distress and approached gently.

"Everything okay, Sophia?" he asked.

Sophia sighed, gesturing at her notebook filled with crossed-out math problems. "I just can't seem to get these algebra equations right, Mr. Johnson. No matter how many times I try, I keep making mistakes."

Mr. Johnson nodded sympathetically. "I understand, Sophia. Math can be challenging. But what if I told you there's a new tool that might be able to help?"

Sophia looked up, a glimmer of hope in her eyes. "Really? What kind of tool?"

"It's called an AI study buddy," Mr. Johnson explained. "It's like having a personal tutor available whenever you need help. Would you like to give it a try?"

Sophia's eyes widened with curiosity. "An AI study buddy? That sounds amazing! How does it work?"

Mr. Johnson smiled, pulling out his tablet. "Let me show you. This might just change the way you think about learning..."

This scene, playing out in classrooms across the world, marks the beginning of a revolutionary shift in education. The integration of Artificial Intelligence (AI) into learning environments is not just a futuristic concept – it's a present reality that's reshaping how students engage with knowledge, how teachers facilitate learning, and how parents support their children's educational journey.

In this chapter, we'll explore the exciting world of AI in education, with a particular focus on Generative AI. We'll discuss its potential to transform learning experiences, examine real-world applications, and consider both the promises and challenges of this technology. By the end of this chapter, you'll have a solid foundation for understanding how AI can become a powerful ally in your educational journey.

1.1 The AI Revolution in Education

The integration of AI in education represents one of the most significant shifts in learning methodology since the invention of the printing press. Just as books democratized access to knowledge, AI has the potential to personalize and optimize the learning experience for every student.

To understand the impact of AI in education, let's first clarify what we mean by AI. At its core, Artificial Intelligence refers to computer systems that can perform tasks that typically require human reasoning. These tasks include learning, problem-solving, pattern recognition, and language understanding.

In the context of education, AI manifests in various forms:

1. Adaptive Learning Platforms: These systems use AI to adjust the difficulty and style of content based on a student's performance and preferences. For example, Knewton Alta can create a personalized learning path for each student, ensuring they master foundational concepts before moving on to more advanced topics.

2. Intelligent Tutoring Systems: AI-powered tutors can provide one-on-one support, answering questions and offering explanations tailored to each student's level of understanding. Carnegie Learning's MATHia is an excellent example of this technology in action.

3. Automated Grading Systems: AI can assist teachers by automatically grading objective assessments and providing preliminary feedback on essays, freeing up time for more personalized instruction.

4. Learning Analytics: AI can analyze vast amounts of data to identify patterns in student performance, helping educators make data-driven decisions about curriculum and teaching strategies.

5. Accessibility Tools: AI-powered tools can make education more accessible for students with disabilities, such as text-to-speech technology for visually impaired students or speech recognition for those with mobility impairments.

The potential of AI in education goes beyond just improving academic performance. It has the power to foster curiosity, encourage critical thinking, and prepare students for a future where AI will be an integral part of many professions.

Consider the case of Emily, a 7th-grade student who once struggled with math. Traditional teaching methods left her feeling frustrated and behind her peers. However, when her school introduced an AI-driven math tutoring program called Thinkster Math, Emily's relationship with mathematics transformed dramatically.

Thinkster Math used AI to analyze Emily's problem-solving process, identifying specific areas where she struggled. It then generated personalized exercises and provided real-time feedback, adapting its approach based on Emily's progress. The immediate, non-judgmental feedback helped Emily build confidence, and the personalized content kept her engaged.

Within a few months, Emily's math grades improved significantly. More importantly, she began to see math not as a dreaded chore, but as an exciting challenge. This shift in perspective is a prime example of how AI can not only improve academic performance but also change students' attitudes towards learning.

1.2 The Power and Promise of Generative AI

While adaptive learning platforms and intelligent tutoring systems represent significant advancements in educational technology, the emergence of Generative AI marks a quantum leap in the potential of AI in education.

Generative AI refers to AI systems that can create new content, rather than simply analyzing or responding to existing information. These systems use advanced machine learning techniques, particularly deep learning and neural networks, to generate human-like text, images, or even create code.

The most prominent examples of Generative AI are large language models like GPT (Generative Pre-trained Transformer) developed by OpenAI, and similar models created by companies like Anthropic, and Google. These models have been trained on vast amounts of text data, allowing them to understand and generate human-like text across a wide range of topics and styles.

In the context of education, Generative AI opens up exciting new possibilities:

1. Personalized Content Creation: Generative AI can create custom learning materials tailored to a student's interests, learning style, and current knowledge level. For instance, if a student is struggling with fractions but loves basketball, the AI could generate math problems that use basketball statistics, making the content more engaging and relevant.

2. Writing Assistance: AI-powered writing tools like Quill and NoRedInk can help students brainstorm ideas, structure their essays, and refine their writing style. These tools don't just correct grammar; they can provide suggestions for improving clarity, coherence, and stylistic elements.

3. Language Learning: Generative AI can create infinite practice conversations for language learners, adapting to their proficiency level and focusing on areas where they need the most practice.

4. Problem Generation: In subjects like math and science, Generative AI can create an endless supply of practice problems, each slightly different, allowing students to master concepts through varied repetition.

5. Explanations and Summaries: If a student doesn't understand a concept, Generative AI can provide alternative explanations, breaking down complex ideas into simpler terms or explaining them from different angles until the student grasps the concept.

6. Creativity Support: In subjects like creative writing or visual arts, Generative AI can serve as a brainstorming partner, suggesting plot ideas, character traits, or visual concepts that students can use as springboards for their own creativity.

Let's look at an example of how Generative AI can enhance education. Michael, a parent of a high school student passionate about creative writing, discovered an AI-powered writing assistant called GPT-4o. Together, Michael and his daughter explored how this tool could support her writing process.

They found that GPT-4o could generate story prompts, suggest plot twists, and even help develop character backstories. However, the real value came not from using the AI's output directly, but from how it sparked new ideas and pushed the boundaries of his daughter's creativity.

For instance, when stuck on developing an antagonist for her story, Michael's daughter asked GPT-4o to generate a character description. The AI came up with a complex character with conflicting motivations. While she didn't use this character exactly as the AI described, it inspired her to think more deeply about character motivations and moral ambiguity in her own writing.

Michael observed that interacting with the AI became a form of collaborative brainstorming. It challenged his daughter to think critically about the suggestions it made, deciding what to keep, what to modify, and what to discard. This process not only improved her writing but also honed her critical thinking and decision-making skills.

1.3 The Current Landscape of AI in Education

As we explore the potential of AI in education, we need to understand the current state of AI integration in schools and homes. While the examples we've discussed showcase the exciting possibilities of AI, the reality is that widespread adoption is still in its early stages.

Many schools are just beginning to experiment with AI-powered tools, often starting with specific applications like adaptive math programs or writing assistance software. The level of AI integration can vary significantly between schools, often depending on factors like funding, technological infrastructure, and staff training.

At home, the use of AI for educational purposes is growing, particularly in the wake of the global shift towards remote learning during the COVID-19 pandemic. Parents and students are increasingly turning to AI-powered tutoring apps, language learning programs, and educational games to supplement traditional learning methods.

However, it's crucial to note that the AI landscape in education is rapidly evolving. New tools and applications are constantly emerging, and existing ones are being refined and improved. This dynamic environment offers exciting opportunities, but it also presents challenges in terms of keeping up with the latest developments and ensuring responsible use of these powerful tools.

1.4 Potential Benefits and Challenges of Using AI As a Study Aid

As with any transformative technology, the integration of AI in education comes with both significant benefits and important challenges to consider.

Benefits:

1. Personalization: AI can tailor the learning experience to each student's individual needs, pace, and learning style.

2. Immediate Feedback: AI tools can provide instant feedback on assignments, allowing students to learn from their mistakes in real time.

3. 24/7 Availability: Unlike human tutors, AI study aids are available anytime, allowing students to learn at their own convenience.

4. Engagement: Interactive AI tools can make learning more engaging and enjoyable, potentially increasing student motivation.

5. Data-Driven Insights: AI can provide detailed analytics on student performance, helping both learners and educators identify areas for improvement.

6. Accessibility: AI can make education more accessible for students with disabilities or those in remote areas with limited access to educational resources.

Challenges:

1. Digital Divide: Not all students have equal access to the technology required to benefit from AI-powered learning tools.

2. Privacy Concerns: The use of AI in education involves collecting and analyzing large amounts of student data, raising important privacy considerations.

3. Overreliance on Technology: There's a risk that students might become too dependent on AI tools, potentially hindering the development of independent thinking and problem-solving skills.

4. Quality and Reliability of AI-Generated Content: While impressive, AI-generated content can sometimes contain errors or biases that students need to learn to identify and critically evaluate.

5. Integration with Traditional Education: Finding the right balance between AI-assisted learning and traditional teaching methods can be challenging for educators and institutions.

6. Ethical Considerations: The use of AI in education raises ethical questions about fairness, accountability, and the role of technology in human learning and development.

1.5 What Students and Parents Can Expect from Generative AI in Education

As we embark on this journey of exploring Generative AI in education, it's important to set realistic expectations and prepare for the exciting possibilities ahead.

For Students:

1. More Engaging Learning Experiences: Expect learning to become more interactive and tailored to your interests and learning style.

2. Personalized Support: Look forward to having a tireless study buddy that can offer help and explanations whenever you need them.

3. New Ways of Thinking: Prepare to engage with AI in ways that challenge you to think critically and creatively.

4. Skill Development for the Future: Anticipate gaining valuable experience in working with AI, a skill that will be increasingly important in many future careers.

5. Potential Frustrations: Be prepared for occasional limitations or misunderstandings from AI tools. Remember, they're powerful but not perfect.

For Parents:

1. A New Role in Your Child's Education: Expect to take on the role of a guide, helping your child navigate the world of AI-assisted learning.

2. Opportunities for Collaboration: Look forward to new ways of engaging with your child's education, such as exploring AI tools together.

3. Potential Concerns: Be prepared to address new challenges, such as ensuring responsible AI use and maintaining a healthy balance with traditional learning methods.

4. Continuous Learning: Anticipate the need to stay informed about AI developments in education to effectively support your child.

5. Exciting Possibilities: Look forward to seeing your child engage with learning in new and exciting ways, potentially discovering passions and talents that AI tools help uncover.

1.6 Embarking on Your AI Adventure: A Practical Exercise

Now that we've explored the landscape of AI in education, it's time to take your first steps into this exciting new world. This exercise is designed for students and parents to complete together, fostering collaboration and shared understanding.

Exercise: Exploring AI-Powered Learning Tools

1. Research and Exploration:

   Visit the websites of the following AI-powered learning tools:

   - Thinkster Math: https://hellothinkster.com/

   - Quill: https://www.quill.org/

   - Knewton Alta: https://www.knewton.com/

   - NoRedInk: https://www.noredink.com/

   - Querium: https://www.querium.com/

   - Labster: https://www.labster.com/

2. Selection:

   Choose one or two tools that align with the student's current learning needs and interests. If possible, create trial accounts to explore the features more deeply.

3. Hands-On Experience:

    Have the student complete a lesson, assignment, or activity using the chosen tool(s). Parents should observe and engage in the process, asking questions and encouraging the student to think aloud as they navigate the platform.

4. Reflection and Discussion:

    After the activity, discuss your experiences together. Consider the following questions:

    - What aspects of the AI-powered tool(s) did you find most helpful or engaging?

    - Were there any features that were confusing or frustrating?

    - How do you think the AI features enhanced the learning experience compared to traditional methods?

    - What improvements or additional features would you like to see in these tools?

    - How do you think tools like this could be integrated into regular study routines?

5. Plan for Integration:

    Based on your experience and discussion, create a plan for how you might incorporate AI-powered learning tools into your existing study routine. Consider:

    - Which subjects or topics could benefit most from AI assistance?

- How often would you use these tools?

- How will you balance AI-assisted learning with traditional methods?

- What goals would you like to achieve using these tools?

6. Ethical Consideration:

    - Discuss any ethical considerations or concerns that arose during your exploration of these tools. For example:

    - How do you ensure that using AI tools enhances learning rather than replacing independent thinking?

    - What privacy considerations should you be aware of when using these tools?

    - How can you verify the accuracy of information provided by AI tools?

7. Reflection Journal:

    Encourage the student to start a reflection journal about their experiences with AI in learning. This can be a valuable tool for tracking progress, noting insights, and identifying areas for further exploration.

Remember, this exercise is just the beginning of your AI adventure in education. As you continue to explore and learn together, you'll discover even more ways to harness the power of Generative AI to support growth and success in learning.

1.7 Looking Ahead: Your Journey with Generative AI in Education

As we conclude this introductory chapter, it's important to remember that the integration of AI in education is an ongoing journey, not a destination. The field is rapidly evolving, with new tools, applications, and ethical considerations emerging regularly.

In the coming chapters of *Generative AI in Education*, we'll delve deeper into various aspects of AI-assisted learning:

- Chapter 2 will address the critical topic of staying safe and smart with AI, exploring ethical considerations and providing guidelines for responsible use.

- Chapter 3 will introduce you to the art and science of crafting effective prompts, a key skill in harnessing the power of Generative AI.

- Chapter 4 will provide a comprehensive overview of popular AI tools like ChatGPT, Claude, and Google's Gemini, offering practical guidance on how to use these tools effectively in educational contexts.

- Subsequent chapters will explore topics such as integrating AI into home learning environments, collaborating with teachers on AI use, tracking progress with AI-assisted learning, and preparing for the future of education in an AI-enhanced world.

As you progress through this book, remember that the goal is not to replace traditional learning methods entirely, but to enhance and supplement them with the power of AI. The most effective approach will likely be a balanced one, combining the best of AI-assisted learning with time-tested educational practices.

Also, keep in mind that while AI can be an incredibly powerful tool, it's ultimately just that – a tool. The real magic happens when curious, motivated learners like you engage with these tools critically and creatively. Your questions, your insights, and your unique perspective are what will truly drive your learning forward.

So, as we embark on this exciting journey together, stay curious, remain open to new possibilities, and don't hesitate to question and critically evaluate the AI tools you encounter. Remember, in the world of AI-assisted learning, you are not just a passive recipient of information – you are an active explorer, a critical thinker, and a co-creator of your student's educational experience.

Are you ready to take the next step in your AI-enhanced learning journey? Let's move on to Chapter 2, where we'll explore how to navigate the ethical considerations of AI in education and ensure that our use of these powerful tools remains safe, responsible, and truly transformative.

CHAPTER 2

Staying Safe and Smart with AI

The classroom buzzed with excitement as Mr. Thompson, the high school English teacher, handed back the latest batch of essays. Among the students, 16-year-old Alex sat nervously, fingers drumming on the desk. As Mr. Thompson placed the paper face-down before Alex, he paused, a mix of concern and curiosity in his eyes.

"Alex, I'd like to speak with you after class," he said quietly before moving on.

Alex's stomach churned. The essay had been challenging, but with the help of an AI writing assistant, Alex had produced what felt like the best piece of writing ever. Now, that feeling of triumph was replaced by a gnawing uncertainty.

As the bell rang and other students filed out, Alex approached Mr. Thompson's desk.

"Alex, your essay was... remarkable," Mr. Thompson began. "The language was sophisticated, the arguments well-structured. But I couldn't help noticing that it didn't quite sound like your usual writing style. Did you use an AI tool to help you write this?"

Alex hesitated, then nodded slowly. "Yes, I did. I used it to help me brainstorm ideas and structure my thoughts. Is that... not allowed?"

P. Narciso, *Generative AI in Education*, https://doi.org/10.1007/979-8-8688-0844-9_2

Mr. Thompson leaned back in his chair, considering his words carefully. "AI can be a powerful tool for learning, Alex. But it's important that we use it ethically and responsibly. Let's talk about how we can make sure you're getting the most out of these tools while still developing your own skills and voice..."

This scenario highlights the complex ethical landscape that students, teachers, and parents must navigate in the age of AI-assisted learning. As AI tools become increasingly sophisticated and accessible, it's crucial that we approach their use with thoughtfulness, responsibility, and a clear ethical framework.

In this chapter, we'll explore the ethical considerations surrounding the use of AI in education. We'll discuss the importance of parental involvement, provide age-appropriate guidelines for students, examine specific ethical dilemmas, and offer practical exercises to help families engage with these important issues. By the end of this chapter, you'll have a solid foundation for guiding the ethical use of AI in your child's educational journey.

2.1 The Importance of Parent Involvement in Guiding the Ethical Use of AI

As parents and educators, we play a crucial role in shaping our children's values and ethical frameworks. This responsibility extends to the digital realm, including the use of AI tools in education. Here's why parental involvement is so important:

1. Bridging the Knowledge Gap: Many parents may feel out of their depth when it comes to AI technology. However, by actively engaging with these tools alongside our children, we can learn together and bridge this knowledge gap.

2. Modeling Ethical Behavior: Children often learn by example. By demonstrating ethical use of AI tools ourselves, we set a positive model for our children to follow.

3. Creating Open Dialogue: Establishing an environment where children feel comfortable discussing their use of AI tools allows us to address potential issues proactively.

4. Balancing Enthusiasm with Caution: While we want to encourage our children's interest in technology, we also need to help them understand its limitations and potential risks.

5. Reinforcing School Policies: By aligning our guidance at home with school policies on AI use, we create a consistent message for our children.

6. Preparing for the Future: As AI becomes increasingly prevalent in various aspects of life, guiding ethical AI use now prepares our children for future challenges.

To effectively guide our children, we need to educate ourselves about AI tools and their implications. This might involve

- Exploring AI tools ourselves to understand their capabilities and limitations

- Staying informed about developments in AI and education through reputable sources

- Engaging in discussions with teachers and other parents about AI use in schools

- Attending workshops or webinars on digital ethics and AI in education

Remember, the goal isn't to become an AI expert overnight, but to develop enough understanding to have meaningful conversations with our children and guide them effectively.

2.2 Age-Appropriate Guidelines for Students in Different Grade Ranges (3–5, 6–8, 9–12)

As children grow and develop, their understanding of technology and ethics evolves. Therefore, our approach to discussing AI ethics should be tailored to their age and cognitive abilities. Here are some guidelines for different age groups:

Grades 3–5 (Ages 8–11):

At this age, children are developing basic digital literacy skills. The focus should be on introducing the concept of AI and fostering healthy habits around technology use.

- Introduce AI as a helpful tool, like a very smart computer program that can assist with learning.

- Emphasize the importance of asking for help from parents or teachers when using any new technology.

- Teach basic digital citizenship concepts, such as being kind online and not sharing personal information.

- Encourage creativity and critical thinking alongside AI use. For example, if using an AI art generator, have children add their own elements to the generated images.

- Start discussions about giving credit to sources, including when ideas come from an AI tool.

Example Activity: Play a game of "Robot or Human?" Show children various tasks (like solving a math problem or writing a story) and have them guess whether a human or an AI (robot) would be better at it. Use this to discuss AI's strengths and limitations.

Grades 6–8 (Ages 11–14):

As children enter adolescence, they can grasp more complex concepts about AI and its implications.

- Delve deeper into how AI works, explaining concepts like machine learning in simple terms.

- Discuss the potential biases in AI systems and why critical thinking is important when using AI-generated information.

- Introduce the concept of digital footprints and data privacy, explaining how AI systems use data.

- Encourage responsible use of AI tools for schoolwork, emphasizing the importance of understanding concepts rather than just getting quick answers.

- Begin conversations about AI ethics, using real-world examples that are relevant to their lives.

Example Activity: Have students use an AI writing assistant to help brainstorm ideas for a story, then write the story themselves. Discuss how they chose which AI suggestions to use and how they developed the ideas further.

Grades 9–12 (Ages 14–18):

High school students can engage with more sophisticated ethical discussions about AI.

- Explore complex ethical issues related to AI, such as algorithmic bias, privacy concerns, and the potential impact of AI on future jobs.

— Discuss the role of AI in various fields they might be interested in for future careers.

— Encourage critical analysis of AI-generated content, including identifying potential biases or inaccuracies.

— Engage in discussions about the broader societal implications of AI, including issues of equity and access.

— Promote the development of AI literacy skills, including understanding basic programming concepts and data analysis.

Example Activity: Organize a debate on an AI-related ethical issue, such as "Should AI-generated art be eligible for art competitions?" Have students research and argue both sides of the issue.

2.3 Specific Examples of Ethical Dilemmas and How to Approach Them

To help make these guidelines more concrete, let's explore some specific ethical dilemmas that students and parents might encounter when using AI tools in education, along with approaches to address them.

Dilemma 1: Plagiarism and Original Thought

Scenario: Your child uses an AI writing assistant to help with an essay, resulting in a paper that contains passages nearly identical to the AI-generated text.

Approach: This situation provides an opportunity to discuss the importance of original thought and proper attribution. Here's how you might address it:

1. Explain the concept of plagiarism and why it's important to use one's own words and ideas.

2.   Discuss how AI can be a tool for inspiration and brainstorming, but shouldn't replace the student's own writing and critical thinking.

3.   Guide your child in revising the essay, using the AI-generated text as a starting point but expressing ideas in their own words.

4.   Teach proper citation methods, including how to cite AI tools as sources when appropriate.

Practical Example:

Parent: "Let's look at this paragraph. Can you explain to me in your own words what it's saying?"

Child: [Explains the concept]

Parent: "Great! Now, how could you write that using your own unique experiences or examples?"

Child: [Offers ideas]

Parent: "Excellent. Let's work on incorporating those ideas into your essay. Remember, the AI is here to help you think, not to think for you."

Dilemma 2: Over-reliance on AI for Problem-Solving

Scenario: You notice your child consistently turning to an AI math tutor for answers without attempting to solve problems independently.

Approach: This scenario allows us to discuss the balance between using AI as a learning aid and developing independent problem-solving skills.

1.   Explain that while AI can be a helpful tool, the goal of math education is to understand concepts and develop problem-solving abilities.

2.   Encourage a step-by-step approach to using AI tutors: attempt the problem independently first, then use AI for hints or to check work.

3. Discuss the importance of understanding the process, not just getting the right answer.

4. Work with your child to identify which parts of math they find challenging and use AI tools to target those areas for improvement.

Practical Example:

Parent: "I've noticed you've been using the AI tutor a lot. How about we try a different approach? Let's work through this problem together without the AI first. Then, we can check our work with the tutor and see if we missed anything."

Dilemma 3: AI and Creative Expression

Scenario: Your child becomes reliant on AI art generators for art class assignments, potentially stifling their own creative development.

Approach: This situation provides an opportunity to discuss the role of AI in creative processes and the value of human creativity.

1. Acknowledge the impressive capabilities of AI art generators while emphasizing the unique value of human creativity and personal expression.

2. Discuss how AI can be used as a tool for inspiration or to learn new techniques, rather than as a replacement for the creative process.

3. Encourage your child to use AI-generated art as a starting point, then modify and expand upon it with their own ideas and techniques.

4. Explore the ethical implications of using AI-generated art in assignments, including proper attribution and the importance of transparency with teachers.

Practical Example:

Parent: "These AI-generated images are fascinating! But I'm curious – what elements would you add to make this artwork uniquely yours? How could you combine the AI's ideas with your own artistic style?"

2.4 Navigating AI Bias and Ensuring Diverse Perspectives

An important ethical consideration when using AI in education is the potential for bias in AI systems. AI models are trained on large datasets which may contain societal biases, potentially perpetuating or amplifying these biases in their outputs. Here's how to address this issue:

1. Educate about AI Bias: Explain to your child that AI systems can reflect and amplify biases present in their training data or programming.

2. Encourage Critical Thinking: Teach your child to question and critically evaluate AI-generated content, rather than accepting it as unbiased truth.

3. Seek Diverse Sources: Encourage the use of multiple sources, including non-AI sources, when researching topics.

4. Discuss Representation: Talk about the importance of diverse representation in AI development teams and training data.

5. Identify Bias Together: Work with your child to spot potential biases in AI-generated content and discuss their implications.

Practical Example:

Parent: "Let's look at these AI-generated images of scientists. What do you notice about who is represented? Are there any groups that seem to be missing? How might this affect people's perceptions of who can be a scientist?"

2.5 Data Privacy and Digital Footprints

As AI tools become more integrated into education, it's crucial to discuss data privacy with your child. Here are some key points to cover:

1. Explain Digital Footprints: Help your child understand that their online activities, including interactions with AI tools, leave a digital trail.

2. Discuss Data Collection: Explain how AI systems collect and use data to improve their performance, and the potential privacy implications of this.

3. Teach Safe Practices: Guide your child in creating strong passwords, being cautious about sharing personal information, and understanding privacy settings on various platforms.

4. Explore Data Rights: Discuss concepts like the right to be forgotten and data portability in age-appropriate terms.

5. Model Good Behavior: Demonstrate responsible data practices in your own use of technology and AI tools.

Practical Example:

Parent: "Before we start using this new AI learning tool, let's take a look at its privacy policy together. What kind of information do they collect? How do they use it? Are we comfortable with this?"

2.6 Balancing AI Assistance with Independent Learning

One of the key challenges in using AI for education is ensuring that it enhances rather than replaces independent learning. Here are some strategies to maintain this balance:

1. Set Clear Boundaries: Establish guidelines for when and how AI tools can be used for schoolwork.

2. Emphasize Process Over Product: Encourage your child to focus on understanding concepts and developing skills, not just getting the right answers or highest grades.

3. Use AI for Reflection: Have your child use AI tools to review and reflect on their work, rather than to produce work from scratch.

4. Encourage Self-Assessment: Teach your child to evaluate their own understanding and progress, rather than relying solely on AI feedback.

5. Combine AI and Traditional Methods: Integrate AI tools with traditional learning methods, such as textbooks, hands-on experiments, and group discussions.

Practical Example:

Parent: "Great job on using the AI tutor to check your math homework. Now, can you explain to me how you arrived at this answer? What concepts did you use?"

2.7 Exercises for Discussing AI Ethics with Your Child

To help reinforce these concepts and encourage ongoing dialogue about AI ethics, try these exercises with your child:

Exercise 1: AI in the News

Find a recent news article about an AI-related ethical issue. Read it together and discuss:

- What are the potential benefits and risks of this AI application?

- Who might be impacted, positively or negatively?

- How could this technology be used more ethically?

Exercise 2: AI-Assisted Creative Writing

Have your child use an AI writing tool to generate a short story prompt. Then, work together to:

- Develop the prompt into a full story, focusing on adding original ideas and personal style

- Discuss how the AI-generated prompt influenced the final story

- Reflect on the balance between AI assistance and human creativity

Exercise 3: Ethical AI Design

Challenge your child to design an AI learning assistant for their favorite subject. Discuss:

– What features would make the AI helpful and engaging?

– How could the AI be designed to encourage independent thinking?

– What ethical considerations should be built into the AI's design?

Exercise 4: AI Bias Detective

Use an AI image generator to create images for various professions (e.g., doctor, teacher, scientist). Analyze the results together:

– Are certain groups over- or under-represented?

– How might these representations impact people's perceptions?

– How could the AI be improved to provide more diverse and inclusive results?

By engaging in these exercises and maintaining open dialogue about AI ethics, we can help our children develop a strong ethical foundation for using AI tools responsibly and effectively in their education and beyond.

As we navigate this new frontier of AI in education, it's important to remember that ethical considerations are not a one-time discussion, but an ongoing conversation. The field of AI is rapidly evolving, and new ethical challenges will undoubtedly emerge. By fostering critical thinking, open dialogue, and a strong ethical framework, we can empower our children to harness the benefits of AI while navigating its challenges responsibly.

In the next chapter, we'll explore the art and science of crafting effective prompts for AI tools, a crucial skill for maximizing the benefits of AI in education while maintaining ethical use. Remember, our goal is not to fear or avoid AI, but to embrace it as a powerful tool for learning and growth, always guided by strong ethical principles.

CHAPTER 3

The Magic of Prompts: Unleashing Generative AI's Potential

The classroom was abuzz with excitement as Ms. Rodriguez, the 8th-grade science teacher, unveiled the day's project. "Today," she announced, "we're going to explore the wonders of the solar system using an AI assistant."

Fourteen-year-old Jamal's hand shot up. "Ms. Rodriguez, how do we make sure the AI gives us good information? My dad says you can't trust everything you read online."

Ms. Rodriguez smiled. "Excellent question, Jamal. The key lies in how we ask the AI for information. We call these questions or instructions "prompts." Crafting effective prompts is a skill, and that's exactly what we're going to learn today."

As the students leaned in, eager to begin their cosmic exploration, Ms. Rodriguez continued, "Think of it like this: if you met an alien who knew everything about the universe but didn't understand human language very well, how would you ask it questions to learn what you want to know? That's the art and science of prompting."

P. Narciso, *Generative AI in Education*, https://doi.org/10.1007/979-8-8688-0844-9_3

As we integrate AI tools into education, the ability to craft effective prompts becomes a crucial skill for students, teachers, and parents alike. In this chapter, we'll explore the magic of prompts and how they can help you and your child unlock the full potential of AI language models. We'll delve into the basics of prompt creation, examine strategies for different subjects and grade levels, and discuss the iterative process of refining prompts. By the end of this chapter, you'll have the skills to transform your interactions with AI from simple queries to rich, educational dialogues.

3.1 Understanding the Basics of Prompts

At its core, a prompt is simply a question or instruction given to an AI to elicit a response. However, the art of crafting effective prompts goes far beyond asking basic questions. To understand prompts, let's break them down into their key components:

1. Input: This is the actual text you provide to the AI. It can range from a simple question ("What is photosynthesis?") to a complex instruction ("Write a 500-word essay comparing and contrasting the American and French Revolutions, focusing on their causes, key events, and lasting impacts.").

2. Context: This is additional information you provide to guide the AI's response. For example, you might specify the grade level, subject area, or desired perspective.

3. Format: This instructs the AI on how to structure its response, such as a list, paragraph, dialogue, or step-by-step explanation.

4. Constraints: These are limitations or specific requirements you set for the AI's response, such as word count, tone, or focus areas.

The quality and specificity of your prompt directly influence the usefulness of the AI's response. Well-crafted prompts lead to informative, targeted, and engaging responses that support learning and understanding. Conversely, vague or poorly structured prompts may result in generic, irrelevant, or even misleading responses.

Consider these two prompts about the same topic:

Prompt 1: "Tell me about the water cycle."

Prompt 2: "Explain the water cycle in simple terms suitable for a 4th-grade science class. Include the main stages of the cycle, how water changes forms, and why this cycle is important for life on Earth. Provide a brief, easy-to-understand example of how the water cycle works in nature."

The second prompt is likely to yield a more useful and age-appropriate response because it provides context (4th-grade level), specifies the desired content (main stages, changes in water forms, importance), and requests an example for better understanding.

3.2 The Art and Science of Crafting Effective Prompts

Creating effective prompts is both an art and a science. It requires an understanding of the learning objectives, the subject matter, and the capabilities of the AI language model being used. When designing prompts, consider the following elements:

1. Clarity: Ensure your prompt is clear, specific, and easy to understand. Avoid ambiguity and vagueness.

2. Context: Provide relevant background information to help the AI understand the scope and intent of your prompt. This can include specifying the subject area, grade level, or learning objectives.

3. Format: Specify the desired format or structure of the response, such as a step-by-step explanation, a compare-and-contrast analysis, or a creative story.

4. Length: Indicate the expected length or depth of the response, such as a brief summary, a detailed explanation, or a comprehensive essay.

5. Examples: If applicable, provide examples of what you're looking for to guide the AI's response and ensure it meets your expectations.

6. Perspective: If relevant, specify the viewpoint or approach you want the AI to take in its response.

7. Engagement: Craft your prompt in a way that encourages critical thinking and creativity, rather than just fact recitation.

8. Ethical Considerations: Be mindful of potential biases and ensure your prompts promote inclusive and respectful learning.

Let's look at an example of how these elements can be incorporated into a prompt:

"As a 7th-grade history teacher, create a lesson plan for a 45-minute class on the Industrial Revolution. Include:

1. A brief (2–3 sentences) introduction to the topic

2. Three main discussion points, each with a key fact and an engaging question for students

3. A simple hands-on activity that demonstrates a concept from the Industrial Revolution

4. A short conclusion that links the Industrial Revolution to modern technology

Ensure the content is age-appropriate and encourages critical thinking about the positive and negative impacts of industrialization."

This prompt provides clear context (7th-grade history), specifies the format and length (45-minute lesson plan with specific components), and encourages engagement and critical thinking.

3.3 Sample Prompts and Exercises for Different Subjects and Grade Levels

To help you put these concepts into practice, let's explore some sample prompts and exercises for different subjects and grade levels:

English Language Arts (Grades 6–8):

Prompt: "Analyze the character development of the protagonist in the novel [insert book title]. Provide three key moments in the story that showcase significant changes in the character's personality, beliefs, or actions. For each moment, explain:

1. What happened in the story

2. How it affected the character

3. Why this change is important to the overall narrative

Conclude with a brief reflection on how these changes contribute to the theme of coming-of-age in the novel. Keep your response suitable for a 7th-grade reading level."

Exercise: Have your child use this prompt to generate a response from the AI. Then, work together to identify additional examples from the text that support or contradict the AI's analysis. Encourage your child to write a short paragraph explaining whether they agree or disagree with the AI's interpretation, using evidence from the book to support their argument.

Mathematics (Grades 3–5):

Prompt: "Explain the concept of fractions to a 4th-grade student who is struggling with the idea. Include:

1. A simple definition of what a fraction is

2. Two real-world examples of fractions in everyday life

3. A step-by-step guide on how to add fractions with like denominators, using visual aids in your explanation

4. A fun, hands-on activity that helps reinforce the concept of fractions

Keep the language simple and engaging, and provide encouragement throughout the explanation."

Exercise: After generating the AI's response, work with your child to create physical representations of the fractions mentioned, using household items or drawings. Have your child teach you about fractions using the AI's explanation as a guide, encouraging them to add their own examples or explanations.

Science (Grades 9–12):

Prompt: "Create a comprehensive study guide on the process of cellular respiration for a 10th-grade biology class. Include:

1. A clear, concise definition of cellular respiration

2. The main stages of cellular respiration (glycolysis, citric acid cycle, electron transport chain) with a brief explanation of each

3. The key molecules involved in the process (e.g., glucose, ATP, oxygen) and their roles

4. A comparison between cellular respiration and photosynthesis, highlighting their interconnectedness

5. Real-world applications or examples of how understanding cellular respiration is important in fields like medicine or environmental science

Provide the information in a mix of paragraphs and bullet points for easy readability. Include 2–3 thought-provoking questions that encourage critical thinking about the topic."

Exercise: Have your child use the AI-generated study guide as a basis for creating a visual representation of cellular respiration, such as a flowchart or infographic. Encourage them to research additional sources to verify and expand upon the AI's explanation, and to prepare answers to the thought-provoking questions provided.

Social Studies (Grades 6–8):

Prompt: "Compare and contrast the ancient civilizations of Egypt and Mesopotamia for a 6th-grade social studies class. In your response, address the following points:

1. Geographic location and how it influenced each civilization's development

2. Major achievements and contributions to human history

3. Social structure and governance

4. Religious beliefs and practices

5. Economic systems and trade

Organize the information in a way that clearly shows the similarities and differences between the two civilizations. Conclude with a brief reflection on how these ancient civilizations have influenced modern society. Keep the language and concepts appropriate for 6th-grade students."

Exercise: Based on the AI's response, have your child create a Venn diagram to visually represent the similarities and differences between Egypt and Mesopotamia. Then, challenge them to imagine they are a time-traveling journalist visiting both civilizations. Ask them to write a short "news article" comparing aspects of life in these ancient societies to modern life, using the information provided by the AI as a starting point.

Foreign Languages (Grades 9–12):

Prompt: "As a Spanish language teacher for 11th-grade students, create a lesson plan for teaching idiomatic expressions related to emotions and mental states. Include

1. A list of 10 common Spanish idioms related to emotions, with their literal translations, actual meanings, and example sentences

2. A brief explanation of why idiomatic expressions are important in language learning

3. Three interactive activities to help students practice and remember these idioms

4. A short dialogue incorporating at least 5 of the idioms, with an English translation

Ensure the content is engaging and appropriate for advanced high school Spanish learners."

Exercise: Have your child use the AI-generated lesson plan to create flashcards for the idioms. Encourage them to practice using the idioms in conversation with you, even if you don't speak Spanish. They can explain the meanings and help you understand how to use them correctly. Consider setting up a weekly "Spanish idiom of the day" challenge to incorporate these expressions into daily life.

3.4 The Iterative Process of Refining and Adapting Prompts

Crafting the perfect prompt often requires an iterative approach. It's a process of continuous refinement based on the AI's responses and the learner's needs. Here's a step-by-step guide to this iterative process:

1. Start with an initial prompt based on your learning objective.

2. Analyze the AI's response: Is it too simple or too complex? Does it address all aspects of your question? Is it engaging?

3. Identify areas for improvement in your prompt.

4. Revise your prompt to address these areas.

5. Test the new prompt and analyze the new response.

6. Repeat steps 2–5 until you achieve the desired result.

Let's look at an example of this process in action:

Initial Prompt: "Explain photosynthesis."

AI Response: [Provides a technical, jargon-heavy explanation suitable for college-level biology]

Analysis: The response is too complex for the intended middle school audience and lacks engagement.

Revised Prompt: "Explain photosynthesis in simple terms for a 7th-grade science class. Include a brief definition, the main ingredients needed, and why it's important for life on Earth. Use an everyday analogy to help students understand the process."

New AI Response: [Provides a simpler explanation with an analogy comparing photosynthesis to a food factory]

Analysis: Better, but could use more structure and interactivity.

Further Revised Prompt: "Create a lesson plan for teaching photosynthesis to 7th-grade students. Include

1. A simple definition of photosynthesis

2. The "ingredients" needed for photosynthesis, explained as a recipe

3. An everyday analogy to help students understand the process

4. A brief explanation of why photosynthesis is crucial for life on Earth

5. A hands-on experiment that demonstrates a concept related to photosynthesis

Keep the language simple and engaging, and include 2–3 check-for-understanding questions throughout the lesson plan."

Final AI Response: [Provides a comprehensive, age-appropriate lesson plan with all requested elements]

This iterative process allows you to fine-tune your prompts to achieve the most effective learning outcomes. It also helps develop critical thinking skills as you and your child learn to analyze and improve upon the AI's responses.

3.5 Challenges and Limitations of AI-Assisted Learning

While AI can be a powerful tool for education, it's important to be aware of its limitations and potential challenges:

1. Accuracy: AI can sometimes provide inaccurate or outdated information. Always verify important facts from reliable sources.

2. Bias: AI models can reflect biases present in their training data. Be critical of perspectives presented and seek diverse viewpoints.

3. Lack of Real-Time Information: Most AI models have knowledge cutoffs and can't provide information about very recent events.

4. Inability to Truly Understand: AI doesn't comprehend information the way humans do. It generates responses based on patterns in its training data.

5. Potential for Over-reliance: There's a risk of students becoming too dependent on AI for answers instead of developing their own critical thinking skills.

6. Privacy Concerns: Be mindful of the personal information shared when using AI tools.

7. Ethical Considerations: Using AI for academic work raises questions about originality and academic integrity.

To address these challenges:

1. Develop Strong Critical Thinking Skills: Teach children to question and verify information, regardless of the source.

2. Use AI as a Tool, Not a Replacement: Emphasize that AI is a learning aid, not a substitute for understanding and original thought.

3. Combine AI with Traditional Methods: Use AI in conjunction with textbooks, human instruction, and hands-on learning.

4. Stay Informed: Keep up with developments in AI education tools and their limitations.

5. Maintain Open Dialogue: Regularly discuss the use of AI tools with your child and their teachers to ensure alignment with educational goals and ethical standards.

3.6 The Importance of Collaboration and Communication

Effective use of AI in education isn't a solitary endeavor. It requires collaboration and open communication among students, parents, and educators. Here's why this is crucial:

1. Shared Understanding: When all parties are on the same page about how AI tools are being used, it ensures consistency in approach and expectations.

2. Ethical Use: Open discussions help establish and maintain ethical guidelines for AI use in academic work.

3. Maximizing Benefits: Sharing strategies and experiences can help everyone learn to use AI tools more effectively.

4. Addressing Concerns: Regular communication allows for prompt addressing of any issues or concerns that arise.

5. Fostering Digital Literacy: Collaborative exploration of AI tools helps develop crucial digital literacy skills for the future.

To promote collaboration and communication:

1. Have Regular Check-ins: Discuss AI use in schoolwork with your child and their teachers.

2. Share Experiences: Encourage your child to share their AI learning experiences with classmates and teachers.

3. Participate in School Initiatives: Attend workshops or information sessions about AI in education offered by your child's school.

4. Create a Supportive Environment: Foster an atmosphere where your child feels comfortable asking questions or expressing concerns about AI use.

5. Engage in Continuous Learning: Stay informed about AI developments and share your learnings with your child and their educators.

3.7 Parent-Student Activity: Prompt Creation Challenge

To reinforce the concepts covered in this chapter, try this engaging activity with your child:

1. Choose a subject your child is currently studying.

2. Set a timer for 15 minutes and challenge each other to create as many effective prompts as possible related to that subject.

3. Review each other's prompts, discussing what makes them effective or how they could be improved.

4. Select the top three prompts and use them with an AI tool like ChatGPT, Claude, or Gemini.

5. Analyze the AI's responses together, considering:

   - How well did the prompt elicit the desired information?

   - Was the response appropriate for your child's grade level?

   - Did the response encourage further thinking or exploration?

6. Work together to refine one of the prompts, then test the improved version.

This activity not only practices prompt creation skills but also encourages collaboration, critical thinking, and iterative improvement.

3.8 Looking Ahead: The Exciting World of Prompt Engineering

As we conclude this chapter, it's important to recognize that what we've covered is just the beginning of the vast and exciting field of prompt engineering. As AI technology continues to advance, the art and science of crafting effective prompts will only grow in importance.

In the next chapter, we'll dive deeper into specific AI tools like ChatGPT, Claude, and Gemini, exploring their unique features and how to leverage them for optimal learning experiences. We'll build on the prompt engineering skills you've developed here.

CHAPTER 4

Your AI Toolkit: ChatGPT, Claude, and Google's Gemini AI

The computer lab hummed with anticipation as Mr. Chen, the high school computer science teacher, stood before his class of eager students. "Today," he announced, "we're going to explore some of the most powerful AI tools available to us as learners. These aren't just fancy chatbots; they're gateways to knowledge and creativity that can revolutionize how we learn and solve problems."

As Mr. Chen pulled up a series of websites on the projector screen, 16-year-old Zoe leaned forward, her eyes sparkling with curiosity. "Are these the same AI tools that my dad uses at work?" she asked.

Mr. Chen nodded, a smile playing at the corners of his mouth. "Indeed, Zoe. But today, we're going to learn how to harness these tools not just for work, but for learning and exploration. By the end of this class, you'll have a toolkit that can help you tackle any subject, from literature to advanced physics."

The students exchanged excited glances as Mr. Chen continued, "Let's start our journey into the world of ChatGPT, Claude, and Google's Gemini AI. Remember, these aren't just tools; they're your new study buddies. And like any good friend, the more you get to know them, the more they can help you grow."

© The Editor(s) (if applicable) and The Author(s), under exclusive license to
APress Media, LLC, part of Springer Nature 2024
P. Narciso, *Generative AI in Education*, https://doi.org/10.1007/979-8-8688-0844-9_4

The integration of advanced AI tools into learning environments is opening up unprecedented opportunities for personalized, engaging, and effective education. In this chapter, we'll explore some of the most powerful and versatile AI tools available: ChatGPT, Claude, and Google's Gemini AI.

We'll delve into how these tools can be leveraged to support learning across various subjects and grade levels, examine their unique features and capabilities, and discuss best practices for their effective and responsible use. By the end of this chapter, you'll have a comprehensive understanding of how to integrate these AI tools into your learning journey, empowering you to explore, create, and problem-solve in ways you might never have imagined.

4.1 Introduction to Generative AI Tools

Before we dive into the specific tools, it's crucial to understand what generative AI is and how it differs from other forms of artificial intelligence. Generative AI refers to AI systems that can create new content, such as text, images, or even music, based on the input they receive. These systems use deep learning algorithms and are trained on vast amounts of data, allowing them to understand patterns and generate human-like responses.

The most well-known examples of generative AI tools are large language models like ChatGPT and Claude, which can engage in conversational interactions and provide information on a wide range of topics. These models have been trained on enormous datasets of text from the Internet, books, and other sources, giving them a broad base of knowledge to draw from.

Key characteristics of generative AI tools include

1. Natural Language Processing: They can understand and generate human-like text, making interactions feel conversational and intuitive.

2. Contextual Understanding: These tools can often grasp context and nuance, allowing for more sophisticated and relevant responses.

3. Multitask Capability: Many generative AI tools can perform a variety of tasks, from answering questions to writing essays to solving complex problems.

4. Continuous Learning: While the core models are fixed, many of these tools are updated regularly to improve their capabilities and knowledge base.

5. Creativity Support: They can assist in creative tasks by generating ideas, suggesting alternatives, or even creating content from scratch.

It's important to note that while these tools are incredibly powerful, they are not infallible. They can make mistakes, reflect biases present in their training data, and sometimes generate plausible-sounding but incorrect information. This is why critical thinking and fact-checking remain crucial skills when using these tools.

Now, let's explore three of the most prominent generative AI tools and how they can be used to enhance learning experiences.

4.2 Using ChatGPT for Education

ChatGPT, developed by OpenAI, is a powerful language model that can understand and generate human-like text. It has taken the Internet by storm, with users marveling at its ability to provide detailed, coherent responses to a wide range of prompts. You can access OpenAI's ChatGPT using this link:

```
https://chatgpt.com/
```

ChatGPT is free with certain restrictions. There is an option to pay for the tools and get access to new features and increased limits on the use of the tool.

So, how can you use ChatGPT to support learning? Let's explore some examples and exercises for different subjects and grade levels.

4.2.1 Examples and Exercises for Different Subjects and Grade Levels

English Language Arts:

- Grades 3–5: Ask ChatGPT to help you brainstorm ideas for a story about a magical adventure. Use its suggestions to write your own unique tale.

- Grades 6–8: Input a passage from a book you're reading and ask ChatGPT to analyze the main character's motivations and actions.

- Grades 9–12: Use ChatGPT to generate writing prompts for different genres, such as persuasive essays, creative writing, or research papers.

Mathematics:

- Grades 3–5: Ask ChatGPT to create word problems involving basic arithmetic operations, and work together to solve them.

- Grades 6–8: Input a math concept you're studying, like fractions or linear equations, and ask ChatGPT to provide explanations and examples.

- Grades 9–12: Use ChatGPT to explore real-world applications of mathematical concepts, such as how calculus is used in physics or engineering.

Science:

- Grades 3–5: Ask ChatGPT to explain the lifecycle of a butterfly or the water cycle in simple terms, and then create a poster illustrating the process.

- Grades 6–8: Input a scientific topic you're curious about, like renewable energy or the human digestive system, and ask ChatGPT to provide an overview and some fun facts.

- Grades 9–12: Use ChatGPT to research and analyze current scientific issues, such as climate change or genetic engineering, and discuss the implications with your family or classmates.

Social Studies:

- Grades 3–5: Ask ChatGPT to tell you a story about a famous historical figure, like Martin Luther King Jr. or Amelia Earhart, and then roleplay an interview with that person.

- Grades 6–8: Input a current event or social issue, and ask ChatGPT to provide different perspectives on the topic. Use this information to engage in a thoughtful discussion.

- Grades 9–12: Use ChatGPT to analyze primary sources, such as historical documents or speeches, and discuss how they reflect the values and beliefs of the time period.

Foreign Languages:

–   Grades 3–5: Ask ChatGPT to translate simple phrases or sentences into the target language, and practice pronouncing them together.

–   Grades 6–8: Input a topic related to the culture or history of a country where the target language is spoken, and ask ChatGPT to provide some interesting facts or stories.

–   Grades 9–12: Use ChatGPT to engage in a conversation in the target language, practicing your communication skills and learning new vocabulary and expressions.

Remember, while ChatGPT is a powerful tool, it's essential to think critically about the information it provides and to fact-check when necessary. Encourage your child to use ChatGPT as a starting point for learning and exploration, but to always verify the information and draw their own conclusions.

Here's an example of what a prompt and response might look like in ChatGPT:

Can you explain the water cycle in simple terms for a 4th-grade student? After your response, provide the student with questions to test that student's understanding and also suggest more areas where the student can gain more understanding of the water cycle.

Sure! Let's start with the explanation of the water cycle in simple terms:

The Water Cycle Explained

The water cycle is like a big circle that water follows on Earth. Here's how it works:

1. **Evaporation**: The sun heats up water from oceans, rivers, and lakes, turning it into water vapor (like steam) that goes up into the air.

2. **Condensation**: The water vapor cools down when it gets high in the sky, turning back into tiny water droplets. These droplets come together to form clouds.

3. **Precipitation**: When the clouds get too heavy with water, the water falls back to Earth as rain, snow, sleet, or hail. This is called precipitation.

4. **Collection**: The water that falls returns to oceans, rivers, and lakes. Some of it soaks into the ground to become groundwater. Then, the cycle starts all over again!

Questions to Test Understanding

1. What happens during evaporation?
2. What forms in the sky when water vapor cools down?
3. What do we call the water that falls from the clouds?
4. Where does the water go after it falls to the ground?
5. Can you name the four main steps of the water cycle?

Suggested Areas for More Understanding

1. **Different Forms of Precipitation**: Learn about rain, snow, sleet, and hail.
2. **Groundwater and Aquifers**: Explore how water soaks into the ground and where it goes.
3. **Importance of the Water Cycle**: Understand why the water cycle is important for plants, animals, and humans.
4. **Role of the Sun in the Water Cycle**: Discover how the sun's energy drives the water cycle.
5. **Cloud Formation**: Learn about different types of clouds and how they form.

These areas will help you understand more about how water moves and changes on our planet. Have fun exploring!

We will go deeper in how to create a prompt in the next section. However, here's a quick introduction on why and how I created this prompt above.

When creating a prompt to explain the water cycle in simple terms for a 4th-grade student, the goal is to ensure the explanation is clear, engaging, and educational. Here's an explanation of why the prompt was created this way and why it is effective.

The primary purpose of the prompt is to provide an easy-to-understand explanation of the water cycle, tailored to the comprehension level of a 4th-grade student. This involves breaking down complex processes into simple steps and using language that is accessible to young learners.

The prompt above was structured into three main parts:

1. Simple Explanation: The first part of the prompt asks for a straightforward explanation of the water cycle. This ensures that the student receives a clear and concise description of the process, using language and examples they can easily relate to.

2. Questions to Test Understanding: The second part requests questions that test the student's understanding. These questions are crucial because they reinforce the key points of the explanation and help the student actively engage with the material. By answering these questions, the student can assess their own comprehension and identify any areas that might need further clarification.

3. Suggestions for Further Exploration: The third part of the prompt encourages the student to explore related topics for a deeper understanding. This fosters curiosity and provides additional learning

opportunities, helping the student to connect
the basic concept of the water cycle to broader
environmental and scientific ideas.

This prompt was effective in that it was clear and simple. By asking
for a simple explanation, the prompt ensures that the response will be
appropriate for a 4th grader's level of understanding. Complex scientific
terms and concepts are avoided or simplified, making the information
accessible. It also included questions to test understanding, making
the learning process interactive. It encourages the student to think
critically about what they have learned and to apply their knowledge.
Lastly, suggesting areas for further exploration motivates the student to
continue learning beyond the initial explanation. This can lead to a more
comprehensive understanding of the water cycle and its importance.

The structure of the prompt follows a logical flow, starting with
an explanation, followed by comprehension checks, and ending with
opportunities for further study. This progression helps to build a solid
foundation of knowledge before expanding into more complex areas.

By designing the prompt in this way, the explanation of the water
cycle becomes an effective educational tool that is both engaging and
informative for a 4th-grade student.

4.2.2 Understanding and Creating Effective Prompts

Let's now dive deep into creating prompts that are effective regardless of
the generative AI tool of your choosing. When using generative AI tools like
ChatGPT, the key to getting the most relevant and useful responses lies in
crafting effective prompts. A prompt is simply the input or question you
provide to the AI tool, which it then uses to generate a response.

To create an effective prompt, consider the following tips:

1. Be specific and clear: Provide enough context and detail in your prompt to help the AI understand exactly what you're looking for. Instead of asking, "What is photosynthesis?" try something like, "Can you explain the process of photosynthesis in simple terms for a 5th-grade science student?"

2. Break down complex topics: If you're dealing with a complex subject or question, try breaking it down into smaller, more manageable parts. This will help the AI provide more focused and relevant responses.

3. Use examples: If you're struggling to articulate your question or idea, try providing an example to help illustrate what you mean. For instance, "Can you provide an example of how to solve a quadratic equation using the quadratic formula, similar to the equation $x^2 + 5x + 6 = 0$?"

4. Specify the format or style: If you have a particular format or style in mind for the AI's response, mention it in your prompt. For example, "Please provide a step-by-step guide to conducting a science experiment to demonstrate the effects of acid rain on plant growth."

5. Iterate and refine: If the AI's response doesn't quite meet your needs, don't hesitate to ask for clarification or provide additional information. You can use the AI's response as a starting point and refine your prompt to get closer to the desired outcome.

As an educator, your approach to creating a prompt will be different. Here's an example of an effective prompt for ChatGPT written by a teacher:

"As a 7th-grade history teacher, create a lesson plan for a 45-minute class on the American Revolution. Include

1. A brief introduction to the topic (2–3 sentences)

2. Three main discussion points, each with a key fact and an engaging question for students

3. A short, hands-on activity that helps students understand a concept from the Revolution

4. A conclusion that relates the American Revolution to modern-day concepts of democracy and freedom

Please ensure the content is age-appropriate and encourages critical thinking about the causes and effects of the Revolution."

This prompt provides clear context (7th-grade level), specifies the topic (American Revolution), and requests a specific format (lesson plan with particular elements). By crafting such a detailed and structured prompt, you're more likely to receive a helpful and informative response from ChatGPT.

We'll now use the subject but from a student's perspective. Here's an example of how you can create an effective prompt in ChatGPT written by a 7th grade student:

"I'm a 7th-grade student learning about the American Revolution. Can you provide a summary of the key events and figures involved in the American Revolution, written in a way that's easy for me to understand? Please also include a few fun facts or interesting stories to help make the information more engaging."

ChatGPT can make mistakes. Check important info.

In this prompt, the student has provided clear context (7th-grade level), specified the topic (American Revolution), and requested a specific format (summary with fun facts and interesting stories). By crafting a detailed and engaging prompt, the student is more likely to receive a helpful and informative response from ChatGPT.

Remember, creating effective prompts is a skill that takes practice. As you guessed, because it is a skill that needs to be learned, I've provided you with practice exercises at the end of this chapter, to help you write effective generative AI prompts. I encourage you and your child to experiment with different prompts and to think critically about the responses received. Over time, you'll both develop a better understanding of how to communicate with AI tools to get the most valuable information and insights.

4.3 Using Claude for Education

Claude is an AI assistant created by Anthropic. You can get access to Claude here:

```
https://claude.ai
```

Like ChatGPT, Claude can engage in conversational interactions and provide information on a wide range of topics. However, Claude has some unique features that make it particularly useful for educational purposes. Let's explore how you can use Claude to support learning.

4.3.1 Examples and Exercises for Different Subjects and Grade Levels

Here's an example of a prompt to explore a 4th grade level history topic using Claude:

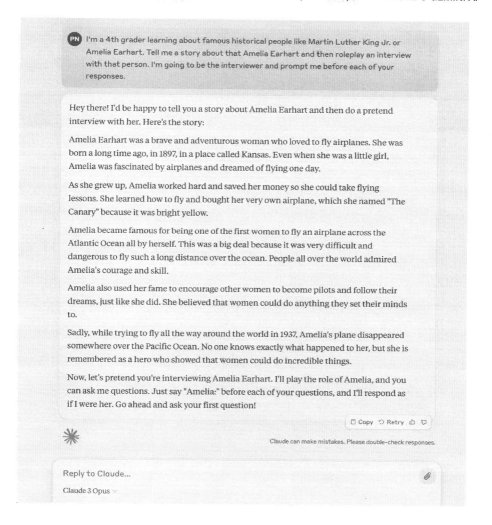

Now let's break down why this prompt works so well:

1. Clear context: The prompt begins by stating, "I'm a 4th grader learning about famous historical people." This provides Claude with a clear understanding of the student's age and the context in which the information will be used, allowing it to tailor its response to the appropriate level.

2. Specific topic: The prompt narrows down the focus to two specific historical figures, Martin Luther King Jr. and Amelia Earhart, rather than asking about famous historical people in general. This helps Claude provide more targeted and relevant information.

3. Engaging format: The student requests a story about Amelia Earhart, followed by a roleplay interview. This format makes the learning experience more engaging and interactive, encouraging the student to actively participate in the conversation and think critically about the information provided.

4. Guided interaction: By stating "I'm going to be the interviewer and prompt me before each of your responses," the student is setting clear expectations for how the roleplay interview will be conducted. This guidance helps Claude understand its role in the conversation and ensures a more structured and coherent interaction.

5. Age-appropriate language: The prompt uses simple, straightforward language that is appropriate for a 4th-grade student. This helps ensure that Claude will respond in a way that is easy for the student to understand and engage with.

By crafting a prompt that provides clear context, focuses on a specific topic, requests an engaging format, sets guidelines for the interaction, and uses age-appropriate language, the student is more likely to receive a response from Claude that is both informative and enjoyable. Additionally, the student is actively engaged with the learning process.

4.4 Using Google AI for Education

Google has developed several AI tools that can be used to support learning, such as Google Lens, which can identify objects and provide information about them, and Google Talk to Books, which allows users to ask questions and receive relevant passages from books. Google's Workspace uses generative AI in their products Docs, Spreadsheets, Gmail, and Slides. All of these tools are powered by Google's generative Ai called Gemini. Let's quickly explore how you can use these tools in your educational journey. I'll be using Gemini to illustrate.

You can access Gemini here:

```
https://gemini.google.com/app
```

Like OpenAI's ChatGPT and Anthropic's Claude, the tool is free to use with limited use. Additionally, there is an option to subscribe to the tool.

4.4.1 Examples and Exercises for Different Subjects and Grade Levels

For this example, we'll use a 4th grade level math problem where the 4th grader might be having trouble understanding fractions and equivalent fractions.

A prompt might be:

> I'm a 4th grader. I'm having trouble understanding what 1/2 and 2/4 mean? Isn't it the same? Can you help me understand and visualize it?

Like our previous examples, we applied the same principles in creating a good prompt by being age appropriate, providing context, and a specific topic. Now let's see Gemini's response below.

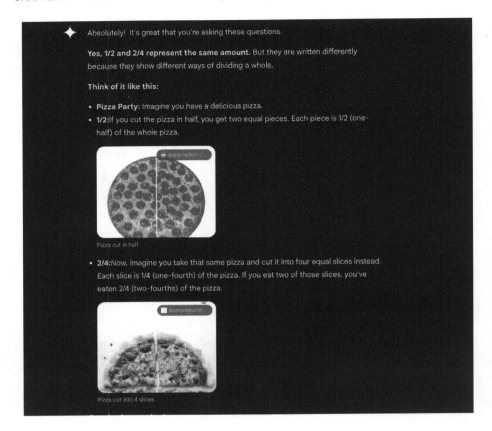

Notice that Gemini is adding visuals to the response? Gemini excels at illustrations and visualizations because of Google's use of sophisticated image generation models.

Let's further explore the rest of Gemini's response to the 4th grade prompt.

See the Connection?

You still ate the same amount of pizza, whether you had one big half-slice or two smaller fourth-slices. That's why 1/2 is equal to 2/4.

Visualizing with Shapes:

1. **Rectangle:**Draw a rectangle. Divide it in half with a line down the middle. Each side is 1/2. Now, draw another line across the middle to make four equal parts. Each part is 1/4. Notice how two of those 1/4 pieces together make up the same amount as one of the 1/2 pieces.

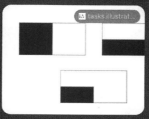

Rectangle cut in half and then in fourths

2. **Circles:**Draw a circle. Divide it in half like you did with the pizza. Each half is 1/2. Now, draw another line across the middle to make four equal pieces. Each piece is 1/4. Again, two of those 1/4 pieces make a 1/2.

Circle cut in half and then in fourths

Important to Remember:

- The bottom number of a fraction (the **denominator**) tells you how many equal parts the whole is divided into.
- The top number (the **numerator**) tells you how many of those parts you have.

Let me know if you'd like more examples or ways to visualize fractions!

We can see that Gemini's response went beyond simple explanations of fractions like 1/2 and 2/4. The inclusion of the "See the Connection?" and "Visualizing with Shapes" sections highlights that Gemini gets smarter as you use it, and it can teach you in a way that works best for you.

By first providing the concrete example of pizza slices, Gemini establishes a familiar context for young learners. The "See the Connection?" section then bridges the gap between the concrete and abstract, helping students solidify their understanding of the equivalence between 1/2 and 2/4.

Gemini also recognizes the importance of visual representations in math education, Gemini offers two additional ways for students to visualize fractions using rectangles and circles. This caters to visual learners and reinforces the concept through different geometric shapes.

While the pizza analogy and visual aids might be most beneficial for younger learners in 3rd or 4th grade, the concepts presented can be easily extended to higher grade levels. For instance, high school students might explore how the concept of equivalent fractions relates to algebraic expressions or geometric proofs.

Lastly, the inclusion of visual representations and open-ended prompts encourages students to actively engage with the material and explore concepts independently. This fosters a deeper understanding of fractions and builds confidence in their math skills.

4.5 Using Microsoft AI for Education

Microsoft has also created AI tools that can assist with learning, such as Microsoft Translator, which can translate text, speech, and images into various languages, and Microsoft Pix, which can help users take better photos and provide information about the objects in the images. Feel free to explore Microsoft's tools on your own for different subjects and grade levels.

4.6 Best Practices and Tips for Utilizing Generative AI Tools

While ChatGPT, Claude, and Gemini are powerful learning aids, it's crucial to use them responsibly and effectively. Here are some best practices to keep in mind:

1. Critical Thinking is Key: Always encourage students to think critically about the information provided by AI tools. These tools can make mistakes or reflect biases, so it's important to verify important information from reliable sources.

2. Use AI as a Starting Point: These tools are great for generating ideas, providing explanations, or offering a different perspective. However, they shouldn't replace independent thinking and problem-solving.

3. Combine AI with Traditional Methods: Use AI tools in conjunction with textbooks, human instruction, and hands-on learning experiences for a well-rounded education.

4. Be Specific in Your Prompts: The more specific and clear your prompts are, the more relevant and useful the AI's responses will be.

5. Iterate and Refine: If you don't get the response you're looking for, try rephrasing your prompt or breaking it down into smaller parts.

6. Encourage Collaboration: Have students discuss the AI-generated content with each other, their teachers, or their parents. This can lead to deeper understanding and critical analysis.

7. Set Clear Guidelines: Establish rules for when and how AI tools can be used for schoolwork. For example, they might be allowed for brainstorming or fact-checking, but not for writing entire essays.

8. Teach Digital Literacy: Help students understand how these AI tools work, including their limitations and potential biases.

9. Focus on Skills, Not Just Content: Use AI tools to help develop skills like critical thinking, problem-solving, and effective communication, not just to acquire factual knowledge.

10. Stay Informed: Keep up with developments in AI education tools and their best practices, as this field is rapidly evolving.

4.7 Potential Limitations and Ethical Considerations

While generative AI tools offer exciting possibilities for education, they also come with limitations and ethical considerations that need to be addressed:

Limitations:

1. Accuracy: AI can sometimes provide inaccurate or outdated information. Always verify important facts from reliable sources.

2. Lack of Deep Understanding: While AI can process and recombine information impressively, it doesn't truly "understand" concepts the way humans do.

3. Limited Context Awareness: AI might not always grasp the full context of a situation or question, leading to irrelevant or inappropriate responses.

4. Potential for Overreliance: There's a risk of students becoming too dependent on AI for answers instead of developing their own critical thinking skills.

Ethical Considerations:

1. Privacy Concerns: Many AI tools collect and process user data. It's important to be aware of what information is being shared and how it's being used.

2. Plagiarism and Academic Integrity: The ease of generating content with AI raises questions about originality and proper attribution in academic work.

3. Bias and Fairness: AI models can reflect and amplify biases present in their training data, potentially perpetuating stereotypes or unfair representations.

4. Access and Equity: Not all students have equal access to AI tools or the technology required to use them, which could create or exacerbate educational inequalities.

5. Transparency: It's not always clear how AI tools arrive at their responses, which can make it difficult to evaluate their reliability or appropriateness.

To address these challenges

1. Develop Strong Critical Thinking Skills: Teach students to question and verify information, regardless of the source.

2. Establish Clear Guidelines: Create and enforce policies on appropriate use of AI tools in academic settings.

3. Promote Digital Literacy: Educate students about how AI works, including its limitations and potential biases. This knowledge will help them use AI tools more effectively and responsibly.

4. Encourage Transparency: When using AI-generated content, students should be open about its use and provide proper attribution.

5. Focus on Process, Not Just Results: Emphasize the importance of understanding the problem-solving process, not just getting the right answer from an AI tool.

6. Discuss Ethical Implications: Engage students in discussions about the ethical considerations of AI use in education and society at large.

7. Ensure Equitable Access: Work toward providing equal access to AI tools and the necessary technology for all students.

4.8 Collaborative Exercises for AI-Assisted Learning

To help students and parents make the most of these AI tools while addressing potential challenges, here are some collaborative exercises you can try:

Exercise 1: Create a Mind Map

1. Choose a topic that your child is currently studying or interested in learning more about.

2. Work together to create a mind map using an AI tool like ChatGPT, Claude, or Gemini.

3. Start by asking the AI to generate a central idea or main topic for the mind map.

4. Then, ask the AI to provide subtopics, key points, and relevant examples to expand the mind map.

5. Discuss the generated information and have your child add their own ideas, questions, or insights to the mind map.

6. Use the completed mind map as a study guide or a starting point for further exploration of the topic.

This exercise helps develop critical thinking skills, encourages active engagement with the material, and teaches students how to organize and synthesize information from multiple sources, including AI.

Exercise 2: AI-Assisted Writing Workshop

1. Have your child choose a writing prompt or assignment related to a subject they are currently studying.

2. Use an AI tool to generate ideas, outlines, or rough drafts for the writing piece.

3. Review the AI-generated content together, discussing its strengths and weaknesses.

4. Have your child write their own version of the assignment, incorporating their own ideas and insights while using the AI-generated content as a guide.

5. Use the AI tool to help with editing and proofreading, discussing any suggested changes or improvements.

This exercise teaches students how to use AI as a brainstorming and editing tool while still emphasizing the importance of original thought and writing skills.

Exercise 3: AI-Powered Flashcards and Quizzes

1. Identify a set of terms, concepts, or facts that your child needs to learn for a particular subject.

2. Use an AI tool to generate definitions, explanations, and examples for each item on the list.

3. Work together to create digital flashcards using a tool like Quizlet or Anki, incorporating the AI-generated content.

4. Have your child study the flashcards, discussing any questions or confusion that may arise.

5. Use the AI tool to generate practice questions or quizzes related to the flashcard content, helping your child reinforce their understanding.

This exercise demonstrates how AI can assist in creating study materials and self-assessment tools, while still requiring active engagement from the student.

Exercise 4: Collaborative Problem-Solving

1. Choose a complex problem or question related to a subject your child is studying.

2. Work together to break down the problem into smaller, more manageable parts.

3. Use an AI tool to generate potential solutions, explanations, or examples for each part of the problem.

4. Discuss the AI-generated content and have your child offer their own ideas or insights.

5. Collaborate to develop a final solution or answer to the problem, incorporating both the AI-generated content and your child's own contributions.

This exercise teaches students how to approach complex problems systematically, use AI as a tool for generating ideas, and critically evaluate and synthesize information from multiple sources.

Exercise 5: AI-Assisted Research Project

1. Have your child choose a research topic related to a subject they are studying.

2. Use an AI tool to generate a list of potential sources, keywords, or research questions related to the topic.

3. Work together to evaluate the relevance and reliability of the AI-generated sources, discussing any potential biases or limitations.

4. Encourage your child to use the AI tool to help summarize or paraphrase information from the sources, while emphasizing the importance of proper citation and avoiding plagiarism.

5. Collaborate to create an outline or draft of the research paper, using the AI-generated content as a starting point and incorporating your child's own analysis and insights.

This exercise teaches valuable research skills, critical evaluation of sources, and proper use of AI tools in academic writing.

4.9 Looking Ahead: The Future of AI in Education

As we conclude this chapter, it's important to recognize that the field of AI in education is rapidly evolving. New tools and capabilities are constantly being developed, and our understanding of how best to integrate AI into learning environments is continually expanding.

Some trends to watch for in the future of AI in education include

1. More Personalized Learning: AI tools may become even better at adapting to individual learning styles and needs, providing truly personalized educational experiences.

2. Enhanced Multimodal Learning: We may see more AI tools that can seamlessly integrate text, images, audio, and video, catering to diverse learning preferences.

3. Improved Natural Language Processing: AI's ability to understand and generate human-like text will likely continue to improve, making interactions with these tools even more natural and intuitive.

4. Greater Integration with Educational Platforms: AI tools may become more tightly integrated with existing educational software and learning management systems.

5. Advancements in AI-Assisted Assessment: We may see more sophisticated AI tools for grading and providing feedback on student work, potentially freeing up more time for teachers to focus on individual student needs.

6. Increased Focus on AI Ethics and Digital Literacy: As AI becomes more prevalent in education, there will likely be a greater emphasis on teaching students about AI ethics, digital literacy, and responsible AI use.

As we navigate this exciting frontier of AI-assisted learning, it's crucial to approach these tools with a balance of enthusiasm and critical thinking. By using AI tools responsibly and effectively, we can enhance the learning experience, foster creativity and critical thinking, and prepare students for a future where AI will play an increasingly important role.

Remember, the goal is not to replace human teaching and learning with AI, but to use AI as a powerful tool to augment and enhance the educational experience. By combining the analytical power of AI with human creativity, empathy, and critical thinking, we can create learning environments that are more engaging, effective, and personalized than ever before.

In the next chapter, we'll explore how to bring AI-assisted learning into the home environment, creating a supportive and enriching space for your child to explore, learn, and grow with the help of these powerful tools.

CHAPTER 5

Bringing AI Home: Enhancing Learning Outside the Classroom

The kitchen table was strewn with textbooks, notebooks, and a tablet propped up against a cereal box. Twelve-year-old Maya sat hunched over her math homework, her forehead creased in concentration. Her mother, Dr. Patel, watched from the doorway, a mix of concern and curiosity on her face.

"Everything okay, sweetie?" Dr. Patel asked, stepping into the room.

Maya looked up, frustration evident in her eyes. "I just can't seem to get these algebra problems right, Mom. I've been staring at them for an hour!"

Dr. Patel sat down beside her daughter, glancing at the tablet screen. "You know, Maya, we have some pretty amazing AI tools that might be able to help. Want to give them a try?"

Maya's eyes lit up with interest. "Really? How would that work?"

"Well," Dr. Patel began, pulling the tablet closer, "let's start by asking our AI assistant to explain the concept in a different way. Then we can work through the problems together, using the AI as our study buddy. What do you say?"

As Maya nodded eagerly, Dr. Patel couldn't help but smile. She knew they were about to embark on a new chapter in Maya's learning journey – one that would bring the power of AI right into their home.

5.1 Setting Up Your AI Learning Headquarters

Creating an effective AI-assisted learning environment at home doesn't require a high-tech setup. Instead, focus on creating a space that's comfortable, conducive to learning, and equipped with the necessary tools. Here's how to get started:

1. Choose a dedicated learning space: Select a quiet area in your home where your child can focus without distractions. This could be a corner of their bedroom, a spot at the kitchen table, or even a converted closet space. The key is consistency and association with learning. For example, you might set up a small desk in the living room corner, decorated with your child's favorite educational posters and a plant to create a calm atmosphere.

2. Ensure reliable Internet access: A stable Internet connection is crucial for accessing AI tools like ChatGPT, Claude, or Gemini. Consider using a wired connection or a Wi-Fi extender if needed to ensure consistent access. If your home has Wi-Fi dead spots, consider setting up a mesh network system to provide comprehensive coverage.

3. Provide appropriate devices: Depending on your child's age and needs, this could be a desktop computer, laptop, tablet, or even a smartphone. Ensure the device is capable of running the AI tools smoothly. For younger children, a tablet with a protective case might be ideal, while older students might benefit from a laptop with a full keyboard for more extensive writing tasks.

4. Organize learning materials: Keep traditional learning materials like textbooks, notebooks, and writing utensils nearby. This allows for easy integration of AI-assisted and traditional learning methods. Consider using color-coded folders or digital organization systems to keep materials for different subjects easily accessible.

5. Create a visually stimulating environment: Decorate the space with educational posters, inspirational quotes, or your child's own artwork to create a positive and motivating atmosphere. You might create a "word wall" with vocabulary related to current topics of study, or display a world map to spark geographical curiosity.

6. Establish a routine: Set up a regular schedule for AI-assisted learning sessions, balancing them with other activities and breaks. For instance, you might dedicate 30 minutes each evening to AI-assisted math practice, followed by 15 minutes of free reading.

7. Ensure proper lighting and ergonomics: Good lighting reduces eye strain and keeps your child alert. Natural light is best, but if that's not possible, use a combination of ambient and task lighting. Also, ensure your child's workstation is ergonomically set up to prevent discomfort during long study sessions.

8. Minimize distractions: Use tools like website blockers or app timers to limit access to potentially distracting content during study time. You might also consider noise-cancelling headphones if your home environment is particularly noisy.

9. Incorporate interactive elements: Include a small whiteboard or chalkboard for brainstorming or working out problems. This can be especially useful when using AI tools to generate practice questions or explain concepts.

10. Create a digital resource library: Set up bookmarks or a dedicated folder on the learning device with links to frequently used AI tools, educational websites, and digital textbooks for easy access.

Remember, the goal is to create a space that your child associates with focused, enjoyable learning experiences. Involve your child in the setup process to ensure the space feels personal and inviting to them.

5.2 Catering to Different Learning Preferences with AI

While the concept of fixed "learning styles" has been largely debunked by recent research, it's true that children often have preferences for how they engage with new information. AI tools offer a unique opportunity to cater to these preferences and make learning more engaging and effective. Here's how:

1. Visual learners: Use AI tools to generate diagrams, infographics, or mind maps to illustrate complex concepts. For example, when studying the water cycle, you could ask ChatGPT to describe each stage, then use an AI image generator to create a visual representation. You might also use AI to generate color-coded notes or visual timelines for historical events.

2. Auditory learners: Utilize text-to-speech features or AI-generated audio explanations. You could have Claude generate a script explaining a scientific concept, then use a text-to-speech tool to turn it into an audio lesson. For language learning, use AI to generate conversations or stories in the target language, which can be listened to and repeated.

3. Kinesthetic learners: Incorporate AI-suggested hands-on activities or experiments. Ask Gemini to propose interactive exercises that reinforce learning through physical engagement. For instance, when learning about geometry, the AI might suggest creating 3D shapes with household items or using body movements to understand angles.

4. Reading/writing preference: Leverage AI tools to generate written summaries, essay outlines, or practice questions for various subjects. For a history lesson, you might ask the AI to create a series of short passages about key events, followed by comprehension questions. Or use AI to help brainstorm and structure essays on complex topics.

5. Multimodal approach: Combine different formats to create a rich, multi-sensory learning experience. For instance, when studying a Shakespeare play, use AI to generate a written summary, create a visual character map, suggest a role-playing activity, and provide audio pronunciations of difficult words.

6. Logical-mathematical thinkers: Use AI to generate problem-solving scenarios or logic puzzles related to the subject matter. For example, when learning about ecosystems, the AI could create a series of "if-then" scenarios about how changes in one part of the ecosystem affect others.

7. Interpersonal learners: While AI can't replace human interaction, it can facilitate it. Use AI to generate discussion topics or collaborative project ideas that your child can explore with peers or family members.

8. Intrapersonal learners: Employ AI to create reflective prompts or journaling exercises that encourage self-analysis and deeper understanding of the material in relation to personal experiences.

9. Naturalistic learners: For those who learn best by seeing patterns and connections to the natural world, use AI to generate examples of how academic concepts apply in nature or real-world scenarios.

10. Musical learners: While not all subjects lend themselves to musical interpretation, you can use AI to find or create educational songs or rhythmic mnemonics for memorizing information in subjects like history or science.

By offering diverse learning experiences, you can help your child discover which approaches work best for them and develop a more flexible, adaptable learning style. The key is to experiment with different methods and observe which ones lead to better engagement and retention for your child.

5.3 Finding the Perfect Balance: AI and raditional Learning

While AI-assisted learning offers exciting possibilities, it's crucial to maintain a balance with traditional learning methods. Here's how to create a well-rounded approach:

1. Use AI as a supplement: Employ AI tools to reinforce concepts learned through traditional methods, not to replace them entirely. For example, after reading a chapter in a history textbook, use ChatGPT to generate discussion questions or summaries. This helps reinforce the material and provides a different perspective on the content.

2. Encourage critical thinking: Have your child evaluate and question AI-generated content. This develops crucial analytical skills and prevents over-reliance on AI. For instance, when using AI to help with a science project, encourage your child to fact-check the AI's suggestions against reliable sources and consider any potential biases in the information provided.

3. Combine AI with hands-on activities: Use AI to generate ideas for experiments or projects, then carry them out in the real world. This bridges the gap between digital and physical learning experiences. For a biology lesson, you might ask the AI to suggest a simple experiment demonstrating photosynthesis, then help your child set it up and observe the results over time.

4. Practice writing skills: While AI can assist with writing, ensure your child still engages in independent writing exercises to develop their own voice and style. Use AI to brainstorm ideas or provide feedback, but have your child do the actual writing. For example, use AI to generate a list of creative writing prompts, but have your child write the stories entirely on their own.

5. Maintain human interaction: Use AI-generated content as a starting point for discussions with family members, teachers, or study groups. This helps develop communication and social learning skills. After using AI to research a topic, encourage

your child to present what they've learned to the
family during dinner, fostering discussion and
deeper understanding.

6. Set AI-free time: Designate periods for learning
    without AI assistance to ensure your child can
    function independently when necessary. This
    could include dedicated reading time, hands-on art
    projects, or outdoor nature observation sessions.

7. Use AI for personalized practice: Leverage AI to
    generate custom practice problems or quizzes based
    on your child's specific needs, supplementing their
    regular coursework. For math, you might use AI to
    create a series of problems targeting the specific
    concepts your child finds challenging.

8. Integrate AI into traditional study methods: Use AI
    to enhance tried-and-true study techniques. For
    example, use AI to help create flashcards, then have
    your child study them using the traditional spaced
    repetition method.

9. Balance screen time: While AI tools are often
    screen-based, ensure there's a healthy mix of on-
    screen and off-screen learning activities. Use AI to
    generate ideas for physical activities that reinforce
    learning concepts.

10. Teach digital literacy alongside AI use: As you
    incorporate AI into learning, take the opportunity to
    teach broader digital literacy skills, including how
    to evaluate online sources, understand data privacy,
    and use technology responsibly.

By striking this balance, you can harness the benefits of AI while ensuring your child develops a well-rounded set of learning skills. The goal is to use AI as a powerful tool in your child's learning toolkit, not as a replacement for traditional learning methods or critical thinking skills.

5.4 Crafting a Personalized AI-Assisted Learning Plan

Creating a tailored learning plan that incorporates AI tools can significantly enhance your child's educational experience. Here's a step-by-step guide to developing an effective plan:

1. Assess current knowledge and skills: Use AI-generated diagnostic tests or quizzes to identify your child's strengths and areas for improvement across various subjects. For instance, you might use ChatGPT to create a series of math problems covering different topics, then analyze which areas your child struggles with most.

2. Set SMART goals: Work with your child to establish Specific, Measurable, Achievable, Relevant, and Time-bound goals. For example, "Improve multiplication skills for numbers 1-12 within the next month, aiming to achieve 90% accuracy on a timed test." Use AI to help break down larger goals into smaller, manageable milestones.

3. Identify preferred learning methods: Observe how your child responds to different AI-assisted learning activities and traditional methods to determine their preferences. You might use AI to generate a variety of learning activities for a single topic – such

as videos, quizzes, and hands-on experiments for a science concept – and note which ones your child engages with most enthusiastically.

4. Select appropriate AI tools: Choose AI tools that align with your child's goals and learning preferences. This might include general-purpose tools like ChatGPT for generating explanations and practice questions, or subject-specific AI tutors for targeted learning in areas like math or language.

5. Create a schedule: Develop a weekly timetable that balances AI-assisted learning with traditional methods, ensuring variety and preventing burnout. For example:

    •   Monday: 30 minutes AI-assisted math practice, 30 minutes independent reading

    •   Tuesday: 45 minutes AI-generated science experiment, 15 minutes reflection writing

    •   Wednesday: 30 minutes language learning with AI chatbot, 30 minutes offline vocabulary games

    •   Thursday: 45 minutes AI-assisted essay planning and writing, 15 minutes physical exercise

    •   Friday: 30 minutes AI-generated history quiz, 30 minutes group discussion of topics learned

6. Design learning activities: Use AI to generate a mix of exercises, projects, and challenges that target your child's goals and appeal to their interests. For a child interested in space, you might ask the AI to create a series of astronomy-themed math problems or a writing prompt about life on other planets.

7. Implement regular check-ins: Schedule weekly or bi-weekly reviews to assess progress, adjust goals, and address any challenges. Use AI to generate reflection questions for these sessions, such as "What was the most interesting thing you learned this week?" or "Which activity did you find most challenging and why?"

8. Encourage self-reflection: Prompt your child to think about their learning process, what's working well, and what could be improved. You might use AI to create a digital learning journal template with guided questions to help your child reflect on their progress.

9. Celebrate milestones: Use AI to generate creative ways to acknowledge and celebrate your child's achievements along the way. This could be anything from generating a personalized certificate of achievement to suggesting a themed family activity related to what your child has been learning.

10. Adapt and evolve the plan: Regularly reassess and adjust the learning plan based on your child's progress and changing needs. Use AI to analyze trends in your child's performance and suggest modifications to the learning approach.

Remember, this plan should be flexible and adaptable. As your child grows and their needs change, be prepared to adjust the plan accordingly. The beauty of using AI in this process is that it can quickly generate new ideas and approaches as needed, helping you keep the learning experience fresh and engaging.

5.5 Fostering Independence and Self-Directed Learning

One of the most valuable skills you can help your child develop is the ability to direct their own learning. AI tools can be particularly useful in fostering this independence. Here's how:

1. Teach effective prompting: Guide your child in crafting clear, specific prompts for AI tools. This skill will help them extract the most useful information and ideas. Start with simple prompts and gradually increase complexity. For example:

    - Basic: "Explain photosynthesis"

    - Intermediate: "Explain photosynthesis in simple terms for a 5th grader"

    - Advanced: "Create a step-by-step explanation of photosynthesis, including its importance for life on Earth and potential impacts of climate change on this process"

2. Encourage exploration: Use AI to generate lists of related topics or concepts, then allow your child to choose which areas they want to explore further. For instance, when studying ancient civilizations, ask the AI to list various aspects of daily life in ancient Egypt, then let your child pick which ones they find most interesting to research deeper.

3. Develop research skills: Show your child how to use AI tools to find relevant sources, then verify information across multiple platforms. Teach them

85

to ask the AI for reputable sources on a topic, then guide them in cross-referencing this information with established academic or scientific sources.

4. Promote problem-solving: When your child encounters a difficulty, encourage them to brainstorm potential solutions with an AI assistant before seeking help from you. For example, if they're stuck on a math problem, they could ask the AI for hints or to break down the problem-solving process into smaller steps.

5. Support goal-setting: Use AI to help your child break down larger goals into manageable steps, fostering a sense of autonomy and achievement. For a long-term project, like writing a research paper, the AI could help create a timeline with specific milestones and tasks.

6. Facilitate self-assessment: Employ AI-generated quizzes or reflection prompts to help your child evaluate their own progress and understanding. After completing a unit of study, have them use AI to create a self-assessment quiz, then reflect on their performance.

7. Encourage curiosity: Use AI to generate thought-provoking questions about topics your child is studying, encouraging them to dig deeper and ask their own questions. This fosters a love of learning and the habit of looking beyond surface-level information.

8. Teach time management: Use AI to help create study schedules or to-do lists, allowing your child to take control of their learning time. The AI can suggest optimal study periods based on your child's daily routine and energy levels.

9. Foster creative thinking: Encourage your child to use AI as a brainstorming partner for creative projects. They might ask the AI for unusual perspectives on a topic or for ideas on how to approach a project in a unique way.

10. Develop critical analysis skills: Teach your child to evaluate AI-generated content critically. Encourage them to fact-check information, consider potential biases, and form their own opinions based on multiple sources of information. By empowering your child to take charge of their learning journey, you're setting them up for success not just in their current studies, but in their future academic and professional endeavors. The ability to independently seek out information, critically evaluate it, and apply it to solve problems is a skill that will serve them well throughout their lives.

11. Encourage interdisciplinary connections: Use AI to help your child draw connections between different subjects. For example, they could ask the AI how a concept they're learning in science relates to something they've studied in history or art. This promotes a more holistic understanding of knowledge and encourages creative thinking.

12. Develop metacognitive skills: Use AI to generate reflection questions that encourage your child to think about their own learning process. Questions like "How did you approach this problem?" or "What strategy worked best for you in understanding this concept?" can help them become more aware of their own learning strategies.

13. Promote resourcefulness: Teach your child to use AI as one tool among many in their learning toolkit. Encourage them to combine AI assistance with other resources like textbooks, online videos, or discussions with peers and teachers.

14. Foster a growth mindset: Use AI to provide encouragement and reframe challenges as opportunities for growth. When your student faces a difficult task, they could ask the AI for motivational quotes or examples of historical figures who overcame similar challenges.

15. Encourage peer learning: While AI can't replace human interaction, it can facilitate peer learning. Show your child how to use AI to prepare for study groups or to generate discussion topics for peer-to-peer learning sessions.

Remember, the goal is not to make your student dependent on AI, but to use AI as a tool to develop their independent learning skills. As they become more comfortable with these strategies, gradually reduce the level of guidance you provide, allowing them to take more initiative in their learning process.

5.6 Addressing Challenges and Celebrating Successes

As with any learning approach, using AI at home will come with its own set of challenges and triumphs. Here's how to navigate both:

Addressing Challenges:

1. Technology issues: Have a backup plan for when Internet or device problems occur. Keep some offline AI-generated materials on hand, such as printed worksheets or downloaded educational games.

2. Information overload: Teach your child to break down complex topics into smaller, manageable chunks. Use AI to help prioritize information and create structured learning plans.

3. Maintaining focus: Establish clear boundaries for AI use and incorporate regular breaks and physical activities into the learning routine. Use AI to generate ideas for quick, energizing activities between study sessions.

4. Ensuring accuracy: Develop a habit of fact-checking AI-generated information against reliable sources. Teach your child how to cross-reference information and identify reputable sources.

5. Avoiding over-reliance: Regularly assess your child's ability to work independently without AI assistance. Gradually increase the complexity of tasks they perform without AI help.

Celebrating Successes:

1. Milestone tracking: Use AI to generate a visual representation of your child's progress toward their goals. This could be a digital growth chart or a virtual badge system.

2. Creative rewards: Ask an AI to suggest unique, personalized ways to celebrate achievements based on your child's interests. This could range from a special outing to a custom-designed certificate.

3. Showcase learning: Create opportunities for your child to demonstrate their new knowledge or skills, perhaps through AI-assisted presentations or projects. This could be a family "show and tell" night or a video presentation shared with relatives.

4. Reflect on growth: Regularly discuss with your child how their learning has evolved and improved with the integration of AI tools. Use AI to generate reflection prompts that encourage your child to think about their learning journey.

5. Share success stories: Encourage your child to share their positive experiences with AI-assisted learning with friends, family, or teachers, fostering a sense of pride and accomplishment.

6. Create a digital portfolio: Use AI to help organize and present your child's best work in a digital format, allowing them to see their progress over time.

7. Set up challenges: Use AI to create personalized learning challenges based on your child's interests and goals. Completing these challenges can be a cause for celebration.

8. Encourage peer recognition: If appropriate, create opportunities for your child to share their AI-assisted projects or achievements with peers, fostering a sense of community and shared learning.

9. Celebrate effort, not just results: Use AI to track and recognize your child's effort and persistence, not just their final achievements. This helps foster a growth mindset.

10. Plan AI-themed celebrations: For major milestones, consider planning celebrations that incorporate AI in fun ways, such as an AI-generated scavenger hunt or a family game night with AI-created trivia questions.

By proactively addressing challenges and enthusiastically celebrating successes, you create a positive, resilient learning environment that prepares your child for the AI-integrated future that awaits them.

As we conclude this chapter, remember that bringing AI into your home learning environment is an exciting journey of discovery for both you and your child. Embrace the process, stay curious, and don't be afraid to experiment with different approaches. With patience, creativity, and the powerful assistance of AI tools, you're well on your way to creating a dynamic, personalized learning experience that will serve your child well into the future.

The key is to view AI as a tool that enhances, rather than replaces, the learning process. By combining AI assistance with your guidance, your child's natural curiosity, and traditional learning methods, you can create a rich, engaging educational experience that prepares your child not just for academic success, but for a lifetime of learning and growth in an increasingly AI-integrated world.

CHAPTER 6

Teaming Up with Teachers: AI in the Classroom

Ms. Johnson stood at the front of the classroom, her eyes sparkling with excitement as she addressed her 7th-grade students and their parents at the annual back-to-school night.

"This year," she announced, "we're embarking on an exciting journey. We're going to explore how AI can enhance our learning experiences right here in the classroom."

A ripple of murmurs spread through the room. Some parents leaned forward, intrigued, while others exchanged uncertain glances.

One parent, Mr. Ramirez, raised his hand. "Ms. Johnson, my daughter Sarah has been using AI tools at home for her studies. How can we help support this initiative in the classroom?"

Ms. Johnson smiled warmly. "That's exactly the kind of collaboration we're hoping for, Mr. Ramirez. Let's talk about how we can work together to make this a success for all our students."

© The Editor(s) (if applicable) and The Author(s), under exclusive license to
APress Media, LLC, part of Springer Nature 2024
P. Narciso, *Generative AI in Education*, https://doi.org/10.1007/979-8-8688-0844-9_6

6.1 Starting the Conversation: Talking to Teachers About AI

Initiating a dialogue with teachers about integrating AI into the classroom is a crucial first step. To approach this conversation effectively, start by scheduling a dedicated meeting with your child's teacher. This shows that you value their time and are serious about the topic. When you meet, begin by sharing your child's positive experiences with AI tools at home. For example, you might say, "Using ChatGPT has really helped Alex understand complex science concepts. He's able to ask questions and get explanations tailored to his level of understanding."

Next, inquire about the teacher's familiarity with AI tools and any current use in the classroom. This helps you gauge their knowledge level and attitude toward AI. Be prepared to offer resources, such as articles, tutorials, or case studies that showcase successful AI integration in education. For instance, you could share a study on how AI-powered adaptive learning improved math scores in a similar school district.

It's important to be open to hearing and addressing any concerns the teacher might have. They might worry about cheating, over-reliance on technology, or the need for additional training. Listen attentively and be ready to discuss these issues constructively. Consider suggesting a pilot program, starting small with AI integration in one subject area. This allows for controlled experimentation and evaluation.

Throughout the conversation, express your willingness to assist in the AI integration process. This could involve providing resources, volunteering time, or connecting the teacher with other AI-savvy parents or professionals. Remember, the goal is to foster a collaborative relationship. Approach the conversation with respect for the teacher's expertise and an openness to their perspective.

6.2 Aligning AI with Classroom Curriculum and Goals

Ensuring that AI tools complement and enhance the existing curriculum is essential for successful integration. Start by sitting down with the teacher to go through the curriculum goals for the year. Identify specific areas where AI could support or extend learning objectives. For each relevant learning outcome, brainstorm how AI tools could be used. For example, if one goal is to improve essay writing, discuss how AI could assist with brainstorming, outlining, or providing feedback on drafts.

Collaborate with the teacher to create sample lesson plans that incorporate AI tools. For instance, in a history class studying ancient civilizations, students could use AI to generate "interviews" with historical figures, then fact-check and expand on the AI's responses. This not only makes the subject more engaging but also develops critical thinking skills.

Design homework tasks that leverage AI tools effectively. In a literature class, students could use AI to generate alternative endings to a story, then write an analysis comparing the AI's version to the original. This exercise encourages creativity while also developing analytical skills.

Discuss how AI can help provide personalized learning experiences. An AI tool could generate math problems at varying difficulty levels based on each student's current abilities, ensuring that each child is appropriately challenged. Work with the teacher to develop methods for assessing learning outcomes when AI tools are used. This might include evaluating students' ability to critically analyze AI-generated content or their skill in crafting effective AI prompts.

Develop clear guidelines for appropriate AI use in assignments and projects. This could include rules about citing AI assistance and distinguishing between AI-generated and original student work. These ethical considerations are crucial for maintaining academic integrity while embracing new technology.

Plan to meet periodically with the teacher to review the effectiveness of AI integration and make adjustments as needed. This ongoing evaluation ensures that the use of AI continues to serve the students' best interests and aligns with learning objectives. By aligning AI use with specific curriculum goals and carefully planning its implementation, you can help ensure that AI enhances rather than disrupts the learning process.

6.3 Collaborating on AI-Assisted Learning Projects and Activities

Exciting possibilities emerge when students, teachers, and parents collaborate on AI-assisted learning projects. Consider organizing an AI-Enhanced Creative Writing Workshop. In this project, students use AI to generate unique story prompts or character profiles. They then write their own stories based on these prompts, using AI for brainstorming or overcoming writer's block. The class could compile their stories into an anthology, with an introduction explaining the AI-assisted process. This not only develops writing skills but also teaches students how to use AI as a tool for creativity.

A Virtual AI-Powered Science Fair is another engaging project. Students could use AI tools to research cutting-edge scientific topics and generate hypotheses. They would then design experiments with AI assistance, predicting outcomes based on AI analysis of similar studies. The final step would be creating interactive digital presentations, incorporating AI-generated visualizations of their data. This project combines scientific inquiry with digital literacy and presentation skills.

For social studies or current events, consider an AI-Assisted Global Issues Debate. Students would use AI to research complex global issues from multiple perspectives. The AI could help generate counterarguments, encouraging students to think critically about different viewpoints. The class could then hold a debate tournament, with AI assisting in real-time fact-checking. This project develops research skills, critical thinking, and public speaking abilities.

Art classes could benefit from an AI-Inspired Art Exhibition. Students would use AI image generators to create abstract prompts for ~~art~~work. They would then interpret these AI-generated images to create their own art pieces in various mediums. The class could host an art show, displaying both the AI-generated prompts and the students' interpretations. This project encourages creativity and helps students understand the relationship between technology and artistic expression.

For a more immersive experience, consider AI-Enhanced Virtual Field Trips. Students would use AI to research historical sites or natural wonders they can't visit in person. The AI could generate immersive, interactive virtual tours based on this research. Students would then act as "tour guides," presenting their AI-enhanced virtual experiences to the class. This project combines research skills, digital literacy, and presentation abilities while making distant places and concepts more accessible.

Language classes could implement an AI-Powered Language Exchange Program. Students would use AI language models to practice conversations in the language they're learning. They could collaborate with partner schools in other countries, using AI for real-time translation during video calls. As a final project, students could create AI-assisted bilingual stories or presentations. This approach combines language learning with cultural exchange and technology skills.

For math and science classes, an AI-Assisted Mathematical Modeling Challenge could be fascinating. Students would use AI to help understand and create complex mathematical models of real-world phenomena. They could compete in teams to develop the most accurate predictive models for scenarios like climate change or population growth. Final projects would be presented to a panel of judges, with students explaining their use of AI in the modeling process. This project develops advanced mathematical skills while also teaching students about the practical applications of AI in scientific research.

These projects not only make learning more engaging but also help students develop crucial skills in AI literacy, critical thinking, and collaborative problem-solving. They demonstrate how AI can be used as a tool to enhance learning across various subjects, preparing students for a future where AI will play an increasingly important role in many fields.

6.4 Educating and Empowering Teachers

Supporting teachers in their journey to integrate AI into their teaching practices is crucial for successful implementation. One effective strategy is to organize AI Bootcamps. These could be weekend workshops where teachers can immerse themselves in AI tools. Provide hands-on training sessions on popular platforms like ChatGPT, Claude, and Gemini. Include sessions on AI ethics and responsible use in education. These bootcamps not only build technical skills but also help teachers understand the broader implications of AI in education.

Creating a comprehensive AI resource library is another valuable step. Develop a digital repository of AI-related lesson plans, prompts, and activities. Include video tutorials on how to use various AI tools effectively in different subjects. Regularly update the library with new tools and best practices. This resource becomes a living document that teachers can refer to throughout the year, supporting their ongoing learning and experimentation with AI.

Consider establishing an AI mentorship program. Pair tech-savvy teachers or parents with those less familiar with AI. Organize regular meetups where mentors can provide one-on-one guidance and support. Encourage mentees to document their AI learning journey to inspire others. This peer-to-peer learning approach can be particularly effective in building confidence and fostering a collaborative culture around AI integration.

Hosting AI in Education conferences can create a broader platform for learning and sharing. Organize annual or semi-annual events featuring expert speakers and hands-on workshops. Include sessions where teachers can share their successful AI integration stories. Provide opportunities for teachers to network and form collaborative AI projects. These events can energize the teaching community and spark new ideas for AI integration.

To recognize and incentivize teacher efforts in AI integration, consider developing an AI Integration Certification Program. Create a multilevel certification process for teachers to demonstrate their AI proficiency. Offer incentives for teachers who complete certifications, such as professional development credits. Celebrate and showcase certified teachers as AI champions within the school community. This not only motivates teachers to develop their AI skills but also creates a cadre of in-house experts who can support their colleagues.

Launching an AI Innovation Grant Program can provide both the resources and the motivation for teachers to experiment with AI in their classrooms. Establish a fund to support teachers in implementing innovative AI-assisted learning projects. Provide both financial resources and expert guidance to grant recipients. Host an annual showcase where grant recipients can present their projects and outcomes. This not only supports individual teacher initiatives but also helps to build a body of evidence and best practices for AI integration in your school or district.

Creating AI Learning Circles can foster ongoing collaboration and learning among teachers. Facilitate regular meetups where teachers can share experiences, challenges, and successes with AI integration. Encourage cross-disciplinary collaboration to explore AI applications across different subjects. Use these circles to identify common challenges and collectively brainstorm solutions. This peer-supported learning environment can be a powerful driver of continuous improvement and innovation.

By investing in these various forms of teacher education and empowerment, schools can create a culture of innovation and continuous learning that benefits the entire educational community. Remember, the goal is not just to introduce AI tools, but to support teachers in reimagining their teaching practices to leverage AI effectively for enhanced student learning outcomes.

6.5 Advocating for Responsible and Equitable AI in Education

As AI becomes more prevalent in classrooms, it's crucial to ensure its use is both responsible and equitable. Start by promoting universal access. Advocate for school-wide or district-wide licenses for AI educational tools to ensure all students have equal access. Push for initiatives to provide necessary devices (laptops, tablets) to students who don't have them at home. Support programs that offer high-speed Internet access to underserved communities. These efforts help to level the playing field and ensure that the benefits of AI in education are available to all students, regardless of their socioeconomic background.

Developing comprehensive AI ethics guidelines is another crucial step. Work with school administrators to create clear policies on AI use in classrooms. Address issues such as data privacy, consent for AI-assisted learning, and guidelines for AI use in assignments and assessments. Ensure these guidelines are regularly reviewed and updated as AI technology evolves. Having a robust ethical framework in place helps to build trust among students, parents, and educators, and ensures that AI is used in ways that align with educational values and goals.

Championing inclusive AI is essential for equitable education. Advocate for the use of AI tools that support multiple languages and diverse cultural perspectives. Push for regular audits of AI tools to identify and address potential biases. Support initiatives to increase diversity in AI development teams to ensure tools are designed with all students in mind. By promoting inclusive AI, we can help ensure that AI-assisted education supports and empowers all learners, rather than perpetuating existing inequalities.

Promoting AI literacy is crucial in preparing students for an AI-driven future. Advocate for the inclusion of AI literacy in the curriculum, starting from elementary grades. Support programs that teach students not just how to use AI, but how to understand its implications, limitations, and potential biases. Encourage critical thinking about AI through debates, essay contests, or AI-focused school projects. By developing AI literacy, we empower students to be informed, critical users and shapers of AI technology.

Supporting ongoing research into the effects of AI in education is vital. Advocate for partnerships between schools and universities to study the long-term impacts of AI in education. Push for transparent reporting of AI use and its effects on student outcomes. Encourage the sharing of best practices and lessons learned across school districts. This research-based approach ensures that AI integration in education is guided by evidence rather than hype or fear.

Fostering community engagement around AI in education helps to build understanding and support. Organize town halls or community meetings to discuss AI in education. Create parent-teacher committees focused on guiding AI integration in schools. Develop outreach programs to educate the broader community about AI in education. By involving the wider community, we can harness diverse perspectives and build a supportive ecosystem for AI-enhanced learning.

Advocating for teacher training and support is crucial for successful AI integration. Push for allocation of resources for ongoing teacher training in AI tools and pedagogies. Support initiatives that provide teachers with time and resources to experiment with AI in their classrooms. Advocate for the creation of AI specialist positions in schools to support teachers and students. By investing in teacher development, we ensure that educators are confident and competent in leveraging AI to enhance their teaching.

Promoting balanced use of AI in education is important to maintain the human element of teaching and learning. Advocate for guidelines that ensure AI complements rather than replaces human teaching. Support initiatives that combine AI-assisted learning with traditional teaching methods and hands-on experiences. Encourage the development of assessment methods that can effectively evaluate learning in AI-enhanced environments. This balanced approach ensures that we harness the benefits of AI while preserving the irreplaceable human aspects of education.

By actively advocating for responsible and equitable AI use in education, we can help ensure that AI enhances learning opportunities for all students, regardless of their background or circumstances. Remember, the goal is not just to introduce AI into classrooms, but to do so in a way that supports our broader educational and societal values.

6.6 Fostering a Collaborative AI Ecosystem in Education

The integration of AI in education is not a solitary endeavor but a collaborative effort that involves students, teachers, parents, school boards, and administrators. Each group brings unique perspectives and concerns to the table, making open communication and mutual understanding crucial for successful implementation.

For educators and administrators, initiating discussions with parents about AI in the classroom requires a thoughtful approach. Begin by organizing informational sessions that introduce parents to the concept of AI in education. These sessions should cover the basics of AI technology, its potential benefits in learning, and address common concerns. Use real-world examples and demonstrations to illustrate how AI tools can enhance the learning experience.

It's important to frame AI as a supplement to, rather than a replacement for, traditional teaching methods. Emphasize how AI can provide personalized learning experiences, offer immediate feedback, and free up teachers' time for more one-on-one interactions with students. Share success stories from other schools or pilot programs within your own institution to provide concrete evidence of AI's positive impact.

Addressing parental concerns head-on is crucial. Many parents may worry about data privacy, screen time, or the potential for AI to replace critical thinking skills. Be prepared with clear policies on data protection and usage. Explain how AI tools are designed to encourage critical thinking rather than replace it, and how their use is balanced with offline activities and human interaction.

Create opportunities for parents to experience AI tools firsthand. Set up demo stations during parent-teacher nights or offer virtual workshops where parents can explore educational AI platforms. This hands-on experience can demystify the technology and help parents see its potential benefits more clearly.

Establish regular channels for ongoing communication about AI initiatives. This could include a dedicated section in school newsletters, updates during parent-teacher conferences, or a specific AI in education parent committee. Encourage parents to share their observations and feedback about how AI is impacting their child's learning at home.

Remember that parents can be valuable allies in AI integration. Many may have professional experience with AI or technical skills that could benefit the school's initiatives. Create volunteer opportunities for parents to contribute their expertise, whether through guest lectures, technical support, or participation in AI curriculum development committees.

School boards play a crucial role in determining the appropriate use of AI in the classroom, and their decisions should be informed by community input. To engage the community effectively, school boards can take several approaches.

First, conduct comprehensive surveys to gauge community sentiment about AI in education. These surveys should assess the community's current understanding of AI, their concerns, and their hopes for how it could improve education. Use the results to inform policy decisions and identify areas where more community education is needed.

Organize public forums or town hall meetings dedicated to discussing AI in education. Invite experts in the field to present on the latest developments and potential applications in schools. Ensure these meetings are accessible to all community members, offering multiple time slots and online participation options.

Form an AI in Education Advisory Committee that includes diverse community representation. This committee should include educators, parents, students, local business leaders, and technology experts. Their role would be to research best practices, review proposed AI initiatives, and make recommendations to the school board.

Collaborate with local universities or tech companies to create educational programs about AI for the community. These could include workshops, webinars, or even a community AI literacy course. Such initiatives can help build a shared understanding of AI across the community.

Develop a transparent decision-making process for AI adoption in schools. Clearly communicate the criteria used to evaluate AI tools and programs, and provide regular updates on implementation progress. Consider creating a public-facing dashboard that tracks AI initiatives, their goals, and their outcomes.

Encourage pilot programs and phased implementations of AI tools. This allows for careful evaluation and adjustment based on real-world results and community feedback. Share the results of these pilots widely, including both successes and challenges encountered.

Remember that AI integration in education is an ongoing process that requires continuous evaluation and adjustment. Regularly revisit AI policies and practices, taking into account new technological developments, emerging research on AI in education, and evolving community perspectives.

By fostering a collaborative ecosystem that involves all stakeholders – students, teachers, parents, administrators, and the broader community – schools can ensure that AI is integrated into education in a way that is ethical, effective, and aligned with community values. This collaborative approach not only leads to better outcomes but also builds trust and support for innovative educational practices.

As we conclude this chapter, it's clear that bringing AI into the classroom is a collaborative effort that requires ongoing communication, experimentation, and adaptation. By working together, parents, teachers, administration, the school board and students can create a dynamic, innovative learning environment that prepares students for the AI-driven future while maintaining the essential human elements of education. Embrace this opportunity to be at the forefront of educational innovation, always keeping the focus on enhancing student learning and development.

CHAPTER 7

Tracking Your AI Adventure: Progress and Impact

The computer lab hummed with excitement as Ms. Rodriguez addressed her 8th-grade class. "Alright, everyone," she began, her eyes sparkling with enthusiasm, "we've been using AI tools in our lessons for a few months now. Today, we're going to take a step back and evaluate how it's impacting our learning."

Amidst the sea of curious faces, Jamal raised his hand. "Ms. Rodriguez, how can we tell if the AI is really helping us learn better?"

"Excellent question, Jamal," Ms. Rodriguez replied. "That's exactly what we're going to explore today. We'll be looking at our progress, analyzing the effectiveness of our AI tools, and even designing our own assessment methods. Who's ready to become AI researchers?"

A chorus of eager voices filled the room, marking the beginning of a new phase in their AI learning adventure.

7.1 Evaluating the Effectiveness of AI-Generated Prompts and Feedback

As we delve deeper into the world of AI-assisted learning, it's crucial to develop a critical eye for the quality and effectiveness of AI-generated content. Not all AI-generated prompts and feedback are created equal, and learning to discern their value is an essential skill for both students and parents.

When assessing the effectiveness of AI-generated prompts and feedback, consider several key criteria. First, examine the relevance of the content. Does the prompt or feedback align with your student's learning goals, interests, and current level of understanding? For instance, if your student is studying ancient civilizations and has expressed particular interest in Egyptian culture, effective AI-generated prompts should incorporate elements of Egyptian history, architecture, or daily life.

Clarity is another crucial factor. The prompt or feedback should be clear, concise, and easy for your student to understand and act upon. Vague or overly complex language can hinder learning rather than facilitate it. For example, a clear prompt for a writing assignment might be: "Describe a day in the life of an Egyptian pharaoh, including details about their daily routines, responsibilities, and interactions with others."

Engagement is key to effective learning. Does the AI-generated content capture your student's attention, spark their curiosity, and motivate them to learn and explore further? Engaging prompts often incorporate elements of storytelling, problem-solving, or real-world applications. For instance, instead of simply asking for facts about photosynthesis, an engaging prompt might challenge students to design an experiment to demonstrate the process.

Accuracy is paramount when using AI-generated content for learning. The information provided should be up-to-date, factually correct, and from reliable sources. This is where human oversight becomes crucial. Encourage your student to fact-check AI-generated information against reputable sources, fostering critical thinking skills in the process.

The constructiveness of feedback is another important aspect to consider. Effective feedback should offer specific, actionable suggestions for improvement, rather than just pointing out errors or shortcomings. For example, instead of simply marking a math problem as incorrect, constructive AI feedback might explain the error, provide a step-by-step solution, and offer additional practice problems targeting the specific concept the student struggled with.

Lastly, consider the tone of the AI-generated content. Is it supportive, encouraging, and appropriate for your student's age and learning style? The tone can significantly impact a student's motivation and self-confidence. Positive, growth-oriented language can inspire students to persevere through challenges and view mistakes as learning opportunities.

By regularly evaluating AI-generated prompts and feedback based on these criteria, you can ensure that your student is receiving high-quality support. This process also teaches valuable critical thinking skills, encouraging students to actively engage with and question the information they receive, rather than passively accepting it.

7.2 Addressing Potential Challenges and Pitfalls

As with any educational approach, AI-assisted learning comes with its own set of challenges and potential pitfalls. Being aware of these issues and having strategies to address them is crucial for a successful learning experience.

One common challenge is the risk of overdependence on AI tools. While these tools can be incredibly helpful, overreliance can hinder the development of independent thinking and problem-solving skills. To address this, encourage a balance between AI-assisted and traditional learning methods. Set aside time for independent work, hands-on experiments, and collaborative projects with peers. Emphasize the importance of using AI as a tool to enhance learning, not as a substitute for critical thinking.

For example, if your student is using an AI writing assistant for essays, encourage them to brainstorm ideas independently before turning to the AI for suggestions. Then, have them critically evaluate and build upon the AI's input rather than accepting it wholesale. This approach fosters a healthy relationship with AI tools while developing crucial cognitive skills.

Inconsistent quality of AI-generated content is another potential pitfall. Not all AI tools are created equal, and even advanced systems can sometimes produce irrelevant or inaccurate information. To mitigate this, be selective in choosing AI tools, opting for reputable platforms with a track record of educational success. Additionally, teach your student to cross-reference AI-generated information with other sources, promoting digital literacy and critical evaluation skills.

A lack of personalization can sometimes be an issue with AI tools. While many AI systems offer a degree of customization, they may not always capture the nuances of your student's unique learning style, interests, or needs. To address this, use AI-generated insights as a starting point, but be willing to adapt and customize based on your student's feedback and progress. Regularly discuss with your student how they feel about the AI-assisted learning process, what's working well, and what could be improved.

Technical difficulties can also pose challenges in AI-assisted learning. AI tools may experience glitches, connectivity issues, or compatibility problems. To ensure continuity in your student's learning journey, always have a backup plan. This could include offline learning activities, alternative resources, or even old-fashioned pen-and-paper exercises. Teaching your student to troubleshoot basic technical issues can also be a valuable life skill.

Another potential challenge is the ethical use of AI in education. Issues like data privacy, algorithmic bias, and the appropriate use of AI-generated content in assignments need to be addressed. Discuss these topics openly with your student, establishing clear guidelines for AI use in schoolwork. Teach them about the importance of citing AI assistance when used and the value of original thought and creativity.

By proactively addressing these potential challenges and pitfalls, you can create a more robust and effective AI-assisted learning environment. Remember, the goal is not to avoid challenges entirely, but to use them as opportunities for growth and learning. Each obstacle overcome enhances your student's resilience, problem-solving skills, and understanding of both the capabilities and limitations of AI technology.

7.3 Tracking Student Improvement with AI Assistance

Monitoring and measuring progress is a crucial aspect of any learning journey, and AI-assisted education is no exception. AI tools offer unique opportunities for tracking student improvement, providing data-driven insights that can inform and enhance the learning process.

One of the primary advantages of AI in tracking student progress is its ability to collect and analyze vast amounts of data in real time. Many AI-powered educational platforms can monitor a student's performance across various subjects, identifying patterns, strengths, and areas for improvement. For instance, an AI math tutor might track not only the number of correct answers but also the time taken to solve problems, the types of mistakes made, and even the student's problem-solving approach.

To effectively track student improvement, start by establishing clear, measurable learning goals. These could be based on curriculum standards, personal interests, or areas identified for improvement. For example, a goal might be to improve multiplication skills for numbers 1–12 within a month, achieving 90% accuracy on timed tests.

Once goals are set, use AI tools to create personalized learning paths. Many AI platforms can adapt their content and difficulty level based on a student's performance, ensuring they are always working at an appropriate level of challenge. This adaptive learning approach can significantly accelerate progress by focusing on areas where the student needs the most support.

Regular assessments are key to tracking improvement. AI can generate customized quizzes and tests that target specific skills and knowledge areas. These assessments can be more frequent and less intrusive than traditional testing methods, providing ongoing feedback without causing undue stress. For instance, a language learning AI might incorporate quick vocabulary checks into daily lessons, tracking improvement over time.

Visualization of progress can be a powerful motivator for students. Many AI platforms offer dashboards or progress reports that graphically represent a student's journey. These visual aids can help students see how far they've come and what areas still need work. Parents and teachers can use these visualizations to have data-informed discussions about a student's progress and adjust learning strategies as needed.

AI can also provide valuable insights into learning patterns and behaviors. For example, it might identify that a student performs better on math problems in the morning or struggles with reading comprehension after extended screen time. These insights can inform decisions about study schedules and learning environments.

However, it's important to remember that AI-generated data should not be the sole measure of a student's progress. Combine AI-driven insights with traditional assessment methods, teacher observations, and student self-reflection for a more holistic view of progress. Encourage your student to keep a learning journal, reflecting on their experiences with AI-assisted learning, noting what works well for them, and identifying areas where they feel they need more support.

Regularly discuss the AI-generated progress reports with your student. Use these discussions to celebrate successes, no matter how small, and collaboratively plan strategies for addressing challenges. This process not only tracks improvement but also develops your student's self-awareness and metacognitive skills, empowering them to take ownership of their learning journey.

Remember, the goal of tracking improvement is not just to measure academic progress, but to foster a love of learning and a growth mindset. Use the data and insights provided by AI tools to encourage persistence, celebrate effort, and cultivate a lifelong passion for learning.

7.4 Sample Prompts and Exercises for Monitoring Progress and Assessing AI's Impact

To help you and your student effectively monitor progress and assess the impact of AI-assisted learning, here are some detailed prompts and exercises you can use:

Progress Tracking Journal: Encourage your student to maintain a weekly learning journal. Each entry should include reflections on their AI-assisted learning experiences, noting new concepts learned, challenges faced, and goals for the coming week. Prompt them with questions like: "What was the most interesting thing you learned using AI this week?" "Did you encounter any difficulties? How did you overcome them?" "How do you feel your understanding has improved since last week?"

AI vs. Traditional Learning Comparison: Design an experiment where your student learns a new concept using both AI-assisted methods and traditional methods. For example, they could study one historical event using an AI tutor and another using a textbook. After both sessions, have them take a quiz on each topic and reflect on the learning experience. Discuss which method they found more effective and why.

Skill Progression Challenge: Choose a specific skill (e.g., multiplication, essay writing, vocabulary in a foreign language) and use AI tools to create a series of progressively difficult challenges. Track your student's performance over time, noting improvements in speed, accuracy, and complexity of tasks completed. Create a visual representation of their progress, such as a graph or chart, to make the improvement tangible.

Peer Teaching Exercise: After your student has learned a new concept using AI tools, challenge them to teach that concept to a sibling, friend, or even you. This exercise not only reinforces their learning but also helps assess how well they've understood and internalized the information provided by the AI.

AI-Assisted Project Portfolio: Have your student create a digital portfolio of projects completed with AI assistance throughout the school year. For each project, include a reflection on how AI tools were used, what was learned, and how the project might have been different without AI assistance. At the end of the year, review the portfolio together, discussing growth and the evolving role of AI in their learning.

Feedback Analysis Workshop: Collect examples of feedback received from AI tools over a period of time. Work with your student to analyze this feedback, categorizing it (e.g., grammar suggestions, content improvements, resource recommendations) and discussing its helpfulness. Use this analysis to create a "guide to interpreting AI feedback" that your student can refer to in future learning.

AI Impact Survey: Create a survey for your student to complete monthly, assessing their perceptions of AI's impact on their learning. Include questions about engagement, understanding, confidence, and enjoyment of learning. Track changes in responses over time to gauge the long-term impact of AI-assisted learning.

Learning Style Adaptation Exercise: Use AI tools to present the same information in different formats (e.g., text, audio, visual, interactive). Have your student engage with each format and then discuss which they found most effective. Use this information to tailor future AI-assisted learning experiences to their preferred learning style.

Real-World Application Challenge: Regularly pose real-world problems that require applying concepts learned through AI-assisted methods. For example, if your student has been studying environmental science, challenge them to use their knowledge to design a sustainable garden for your home. This exercise assesses their ability to transfer AI-assisted learning to practical situations.

Metacognitive Reflection Sessions: Schedule monthly "learning about learning" sessions where you and your student discuss their AI-assisted learning strategies. Use prompts like: "How has your approach to using AI tools changed over time?" "What strategies have you developed for verifying AI-generated information?" "How do you decide when to use AI assistance and when to work independently?"

By engaging in these exercises regularly, you and your student can gain valuable insights into the effectiveness of AI-assisted learning, track progress over time, and continually refine your approach to integrating AI tools into the learning process. Remember, the goal is not just to measure academic progress, but to develop critical thinking skills, foster independence, and cultivate a lifelong love of learning.

7.5 Adapting AI Use Based on Student Needs and Preferences

As you track your student's progress and assess the impact of AI on their learning, it's crucial to remain flexible and willing to adapt your approach based on their evolving needs and preferences. Every student is unique, and what works well for one student may not be as effective for another. The key is to view AI integration as an ongoing process of discovery and refinement.

Start by having regular conversations with your student about their AI-assisted learning experiences. Ask open-ended questions like, "What do you enjoy most about using AI tools for learning?" or "Are there any aspects of AI-assisted learning that you find frustrating or challenging?" These discussions can provide valuable insights into your student's preferences and help you identify areas where adjustments might be needed.

Pay attention to patterns in your student's engagement and performance. Do they seem more motivated when using certain types of AI tools or when studying particular subjects with AI assistance? For example, you might notice that your student thrives when using AI-powered interactive simulations for science lessons but struggles with AI writing assistants for essay composition. Use these observations to tailor your AI integration strategy, focusing on tools and approaches that resonate with your student's learning style.

Consider the balance between AI-assisted and traditional learning methods. Some students may benefit from heavy use of AI tools, while others might need more hands-on, experiential learning opportunities. Be prepared to adjust the ratio of AI to non-AI learning activities based on your student's responses and progress.

Experiment with different AI platforms and tools. If one AI tutor or learning assistant isn't producing the desired results, don't hesitate to explore alternatives. The field of educational AI is rapidly evolving, with new tools and platforms emerging regularly. Stay informed about the latest developments and be willing to switch to tools that better meet your student's needs.

Encourage your student to take an active role in shaping their AI-assisted learning experience. As they become more familiar with AI tools, they may develop preferences for certain features or learning approaches. Empower them to voice these preferences and make choices about how they use AI in their studies.

Be attentive to any signs of AI fatigue or overreliance. If your student seems to be losing interest in AI-assisted learning or becoming too dependent on AI tools, it might be time to scale back and reintroduce more traditional learning methods. The goal is to find a healthy balance that supports your student's learning without overwhelming them.

Remember that adaptation may also involve adjusting your own expectations and approach as a parent or learning guide. Stay open to new ideas and be willing to challenge your assumptions about how learning should happen. Your flexibility and open-mindedness will set a positive example for your student and create a more dynamic, responsive learning environment.

Regularly revisit and revise learning goals based on your student's progress and changing interests. AI tools can be particularly helpful in supporting personalized learning paths, so take advantage of this flexibility to keep your student's education aligned with their evolving needs and aspirations.

Finally, don't forget to celebrate the process of adaptation itself. Help your student understand that learning how to learn – including figuring out which tools and methods work best for them – is a valuable skill in its own right. By modeling a flexible, growth-oriented approach to AI-assisted learning, you're helping your student develop adaptability and resilience that will serve them well throughout their educational journey and beyond.

To conclude, tracking your AI adventure is not just about measuring progress – it's about creating a responsive, personalized learning experience that evolves with your student. By carefully evaluating AI tools, addressing challenges, monitoring improvement, and adapting your approach, you can harness the full potential of AI to support your student's unique learning journey. Remember, the most successful AI-assisted learning experiences are those that remain centered on the learner, using technology as a powerful tool for growth, discovery, and empowerment.

CHAPTER 8

The Future of Learning with AI

As the final bell rang at Westfield High School, the bustling hallways gradually quieted. In the computer lab, however, a small group of students remained, huddled around a sleek, holographic display. Their teacher, Ms. Chen, stood nearby, her eyes sparkling with excitement.

"Alright, team," she began, "today we're going to explore how AI might shape our learning experiences in the coming years. Remember, we're not just passive observers in this technological revolution. We're active participants, helping to shape the future of education."

The students leaned in, captivated by the 3D projection of a futuristic classroom. As Ms. Chen began to explain the potential changes on the horizon, it was clear that the future of education with AI was both thrilling and complex, filled with unprecedented opportunities and challenges that would require careful navigation.

This scene, while imaginary, paints a picture of the growing awareness and excitement surrounding the future of AI in education. As we stand on the brink of transformative changes in how we teach and learn, it's crucial to explore the emerging trends, potential long-term benefits and challenges, and the ways we can empower students to thrive in an AI-driven world.

P. Narciso, *Generative AI in Education*, https://doi.org/10.1007/979-8-8688-0844-9_8

8.1 Emerging Trends and Technologies

The landscape of AI in education is evolving rapidly, with new technologies and approaches emerging at an unprecedented pace. One of the most significant trends is the mainstream adoption of AI across all levels of education, from elementary schools to higher education and professional training. This widespread integration is reshaping traditional educational paradigms and opening up new possibilities for personalized, adaptive learning experiences.

Personalized learning journeys, powered by sophisticated AI algorithms, are becoming increasingly prevalent. These systems can analyze individual learning patterns, preferences, and progress to create tailored educational experiences for each student. For instance, adaptive learning platforms can adjust the difficulty and pacing of content in real time based on a student's performance, ensuring that they are always working at an optimal level of challenge. This personalization extends beyond academic content to include recommendations for extracurricular activities, career guidance, and even mental health support, creating a holistic approach to education that addresses the unique needs of each learner.

Another emerging trend is the enhanced support for teachers through AI-powered tools. Rather than replacing educators, AI is becoming an invaluable ally in the classroom. Teachers are benefiting from AI assistants that can handle administrative tasks, grade assignments, and generate detailed reports on student progress. This automation of routine tasks allows educators to focus more on providing personalized guidance, fostering creativity, and nurturing critical thinking skills. AI-driven insights are also helping teachers identify areas where students might be struggling, enabling timely interventions and support.

The concept of skill-centric education is gaining traction as AI reshapes the workforce. Educational institutions are increasingly using AI to analyze labor market trends and predict future skill requirements.

This data-driven approach allows schools and universities to align their curricula with the evolving needs of the job market, ensuring that students are equipped with the skills and knowledge that will be in demand when they graduate. For example, AI systems might recommend the introduction of courses in emerging fields like quantum computing or sustainable energy technologies based on projected industry growth.

Lifelong learning and reskilling have become essential in the face of rapid technological change, and AI is playing a crucial role in facilitating this continuous education. AI-powered platforms are offering flexible, on-demand learning opportunities that allow professionals to acquire new skills or pivot their careers as needed. These platforms can create personalized learning paths based on an individual's current skills, career goals, and learning style, making the process of upskilling or reskilling more efficient and effective.

The integration of immersive technologies like virtual reality (VR) and augmented reality (AR) with AI is opening up new frontiers in experiential learning. AI-driven VR simulations can provide students with hands-on experience in complex or dangerous scenarios, from conducting virtual chemistry experiments to practicing surgical procedures. These immersive experiences, guided by intelligent AI tutors, can significantly enhance understanding and retention of complex concepts.

Natural language processing (NLP) and conversational AI are transforming the way students interact with educational content. AI-powered chatbots and virtual assistants can provide instant, personalized support to students 24/7, answering questions, offering explanations, and even engaging in Socratic dialogues to deepen understanding. These AI tutors can adapt their communication style to suit the learner's preferences and level of understanding, providing a more engaging and effective learning experience.

Data-driven decision-making is becoming increasingly prevalent in educational institutions. AI-powered analytics platforms can process vast amounts of data from various sources – student performance metrics, attendance records, engagement levels, and more – to provide administrators and policymakers with actionable insights. These insights can inform everything from resource allocation and curriculum design to early intervention strategies for at-risk students.

The rise of AI-generated content is another trend that's reshaping education. AI systems can now create educational materials, from textbooks and lesson plans to interactive quizzes and multimedia presentations. This capability allows for the rapid production of up-to-date, customized learning resources that can be tailored to specific educational contexts or individual learner needs. However, this trend also raises important questions about copyright, authorship, and the role of human expertise in content creation.

Ethical considerations and bias mitigation are becoming central concerns as AI becomes more deeply integrated into education. There's a growing awareness of the need to ensure that AI systems are fair, transparent, and free from bias. This includes addressing issues of data privacy, algorithmic fairness, and the potential perpetuation of societal biases through AI systems. Educational institutions and EdTech companies are increasingly focusing on developing ethical guidelines and implementing safeguards to ensure that AI is used responsibly in educational contexts.

8.2 Potential Long-Term Benefits and Challenges

As we look to the future of AI in education, it's important to consider both the potential long-term benefits and the challenges that may arise. One of the most significant benefits is the potential for truly personalized education at scale. AI systems could eventually create highly individualized learning experiences that adapt not just to a student's academic needs but also to their emotional state, learning style, and personal interests. This level of personalization could lead to dramatically improved learning outcomes and student engagement.

Another long-term benefit is the democratization of high-quality education. As AI-powered educational platforms become more sophisticated and accessible, they could help bridge educational gaps across geographic and socioeconomic divides. Students in remote or underserved areas could potentially access world-class educational resources and personalized tutoring through AI systems, leveling the playing field and providing opportunities that were previously out of reach.

AI could also play a crucial role in early identification and intervention for learning disabilities or difficulties. By analyzing patterns in student performance and behavior, AI systems could flag potential issues much earlier than traditional methods, allowing for timely support and intervention. This early detection and support could significantly improve outcomes for students with learning challenges.

The long-term impact of AI on the teaching profession could be profound. While there are concerns about AI replacing teachers, the more likely scenario is a transformation of the teacher's role. Teachers may evolve into learning facilitators and mentors, focusing on developing students' higher-order thinking skills, creativity, and social-emotional learning while AI handles more routine tasks and provides data-driven insights to inform teaching strategies.

However, along with these potential benefits come significant challenges that need to be addressed. One of the primary concerns is the potential widening of the digital divide. As AI-powered education becomes more prevalent, there's a risk that students without access to these technologies could be left behind, exacerbating existing educational inequalities. Ensuring equitable access to AI-enhanced education will be a crucial challenge for policymakers and educators.

Privacy and data security present another major challenge. As AI systems collect and analyze vast amounts of data about students' learning behaviors and personal characteristics, ensuring the protection of this sensitive information becomes paramount. There will need to be robust policies and technological safeguards in place to prevent misuse of student data and to maintain trust in AI-powered educational systems.

The issue of AI bias and fairness in education is a complex challenge that will require ongoing attention. AI systems trained on historical data may perpetuate existing biases in educational systems, potentially disadvantaging certain groups of students. Ensuring that AI systems are fair and unbiased, particularly in high-stakes areas like college admissions or career guidance, will be crucial.

Another challenge lies in maintaining the human element in education as AI becomes more prevalent. While AI can provide personalized instruction and feedback, it may struggle to replicate the empathy, inspiration, and human connection that great teachers provide. Striking the right balance between AI-driven efficiency and human-centered education will be an ongoing challenge.

The rapid pace of technological change also presents a challenge for educational institutions and educators. Keeping up with the latest AI advancements and effectively integrating them into educational practices will require ongoing professional development and potentially significant changes to teacher training programs.

Lastly, there's the challenge of ethical AI development in education. As AI systems take on more significant roles in educational decision-making, from personalized learning paths to college admissions, ensuring that these systems are transparent, accountable, and aligned with educational values and ethics will be crucial.

8.3 Empowering Students to Thrive in an AI-Driven World

As AI continues to reshape the educational landscape and the broader world, it's crucial that we empower students to not just adapt to these changes, but to thrive in an AI-driven future. This empowerment goes beyond teaching students how to use AI tools; it involves fostering a set of skills, attitudes, and ethical understandings that will enable them to navigate and shape the AI-infused world they will inherit.

One key aspect of this empowerment is developing AI literacy. Students need to understand not just how to use AI systems, but also how they work, their capabilities, and their limitations. This includes learning about the basics of machine learning, understanding how AI makes decisions, and recognizing the potential biases and ethical issues associated with AI systems. By demystifying AI, we can help students become informed and critical users of these technologies.

Equally important is fostering creativity and critical thinking skills. While AI excels at processing vast amounts of data and recognizing patterns, human creativity and critical thinking remain crucial. Encouraging students to think outside the box, question assumptions, and approach problems from multiple angles will help them develop skills that complement, rather than compete with, AI capabilities. Project-based learning, design thinking exercises, and collaborative problem-solving activities can all help nurture these essential skills.

Developing strong information literacy skills is another crucial aspect of empowering students in an AI-driven world. With the proliferation of AI-generated content and the increasing sophistication of misinformation, students need to be adept at evaluating the credibility of information sources, fact-checking, and critically analyzing content. Teaching students to question the origin and potential biases of information, whether it comes from human or AI sources, will be essential.

125

Emotional intelligence and social skills will also be increasingly important in an AI-dominated landscape. While AI can process emotional cues to some extent, the nuances of human interaction, empathy, and emotional understanding remain uniquely human capabilities. Encouraging the development of these soft skills through collaborative projects, peer mentoring, and explicit social-emotional learning programs can help students thrive in both their personal and professional lives.

Entrepreneurial thinking and adaptability are qualities that will serve students well in a rapidly changing, AI-influenced job market. Encouraging students to identify problems, develop innovative solutions, and be comfortable with uncertainty can help prepare them for a future where many jobs may be transformed or may not yet exist. Incorporating entrepreneurship education and teaching adaptability as a skill can help foster this mindset.

Ethics education will play a crucial role in empowering students to navigate the complex moral landscape of an AI-driven world. Students should be encouraged to grapple with ethical dilemmas related to AI, such as privacy concerns, the impact of automation on employment, and the potential for AI to exacerbate social inequalities. By developing a strong ethical framework, students will be better equipped to make responsible decisions about the development and use of AI in their future careers and personal lives.

Another important aspect of empowerment is fostering a growth mindset and a love for lifelong learning. In a world where AI is constantly evolving, the ability to continuously learn and adapt will be crucial. Encouraging students to view challenges as opportunities for growth, to embrace failure as a learning experience, and to cultivate curiosity can help them develop the resilience and adaptability needed to thrive in an AI-driven future.

It's also important to provide students with hands-on experience working with AI technologies. This could involve incorporating AI tools into various subjects, from using natural language processing in language arts to leveraging machine learning in science experiments. By giving students practical experience with AI, we can demystify the technology and help them understand its real-world applications and limitations.

Empowering students also means preparing them to be active participants in shaping the future of AI. This involves encouraging them to think critically about the societal implications of AI and to consider their potential role in developing and implementing AI technologies ethically and responsibly. Engaging students in discussions about AI policy, encouraging them to participate in AI ethics hackathons, or involving them in citizen science projects that use AI can all help foster this sense of agency.

Lastly, it's crucial to empower students by ensuring equitable access to AI education and resources. This means working to bridge the digital divide, providing all students, regardless of their background or location, with opportunities to learn about and work with AI technologies. It also involves ensuring that AI educational tools are designed with diversity and inclusion in mind, capable of adapting to different learning styles, cultural contexts, and individual needs.

8.4 Sample Prompts and Exercises for Discussing the Future of AI with Your Child

Engaging children in discussions about the future of AI can help them develop critical thinking skills and prepare them for the AI-driven world they will inherit. Here are some sample prompts and exercises that parents and educators can use to facilitate these conversations:

1. AI Dream School: Ask your child to imagine and describe their ideal school of the future, where AI is fully integrated into the learning experience. What would a typical day look like? How would AI assist

in their learning? What role would teachers play? This exercise can spark creativity and help children envision positive applications of AI in education.

2. AI Ethical Dilemmas: Present your child with hypothetical scenarios involving AI in education and ask them to discuss the ethical implications. For example: "If an AI system could predict with 90% accuracy which students are likely to struggle in a particular subject, should schools use this information to place students in classes? Why or why not?" This can help develop critical thinking about the ethical use of AI.

3. Job Market of the Future: Engage your child in a discussion about how AI might change the job market in the future. Ask them to research and present on a job that might be created or significantly altered by AI in the next 20 years. This can help them start thinking about how to prepare for a changing workforce.

4. AI Debate: Organize a debate on an AI-related topic, such as "Should AI-generated art be eligible for art competitions?" Encourage your child to research and argue both sides of the issue. This can help develop research skills, critical thinking, and the ability to see multiple perspectives on complex issues.

5. Design an AI Education Assistant: Challenge your child to design an AI education assistant. What features would it have? How would it help students learn? What safeguards would be in place to protect

student privacy and ensure fair treatment? This exercise can help children think creatively about the practical applications and challenges of AI in education.

6. AI and Creativity: Have a discussion about whether AI can be truly creative. Show examples of AI-generated art, music, or writing, and ask your child to compare them with human-created works. This can lead to interesting discussions about the nature of creativity and the unique qualities of human intelligence.

7. Fake News Detective: In this exercise, present your child with a mix of real news articles and AI-generated fake news. Challenge them to distinguish between the two and explain their reasoning. This can help develop critical information literacy skills crucial in an age of AI-generated content.

8. AI Time Capsule: Ask your child to create a "time capsule" predicting what AI and education might look like in 50 years. Encourage them to include drawings, written predictions, and even mock-ups of future AI educational tools. This exercise can help children think long-term about the potential impacts of AI.

9. AI and Equality: Engage your child in a discussion about how AI in education could either promote or hinder equality. Ask them to consider issues like access to technology, potential biases in AI systems, and the global impact of AI-powered education. This can help develop awareness of social issues related to AI.

10. Learning from AI: Have your child spend some time interacting with an AI language model like ChatGPT, asking it questions on a topic they're studying. Afterward, discuss the experience. What did they learn? What were the limitations? How might this type of AI be used in future education? This hands-on experience can provide insight into the current state and future potential of AI in learning.

These exercises and prompts can serve as starting points for ongoing conversations about AI and its impact on education and society. By engaging children in these discussions early, we can help them develop the critical thinking skills and ethical awareness needed to navigate and shape the AI-driven future.

8.5 Conclusion: Embracing the AI-Powered Educational Frontier

As we stand on the brink of this new era in education, it's clear that AI will play a transformative role in shaping how we teach and learn. The potential benefits are immense: personalized learning at scale, democratized access to quality education, early intervention for learning difficulties, and the empowerment of teachers with powerful new tools and insights. Yet, we must also be mindful of the challenges, from ensuring equitable access and protecting student privacy to maintaining the crucial human elements of education and addressing potential biases in AI systems.

The future of learning with AI is not a predetermined path, but one that we collectively shape through our decisions, policies, and innovations. It's crucial that all stakeholders – educators, parents, policymakers, technologists, and students themselves – are involved in steering this transformation. We must strive to create an AI-enhanced educational ecosystem that is equitable, ethical, and truly serves the diverse needs of all learners.

As we empower students to thrive in this AI-driven world, we're not just preparing them for the future job market; we're equipping them with the skills, knowledge, and ethical foundation to become active participants in shaping the future of AI itself. By fostering AI literacy, critical thinking, creativity, and a strong ethical compass, we can ensure that the next generation is not just consumers of AI technology, but informed, responsible creators and decision-makers.

The integration of AI in education is not just about technology; it's about reimagining learning itself. It's about creating educational experiences that are more engaging, more effective, and more attuned to the individual needs of each learner. It's about freeing educators from routine tasks so they can focus on what they do best: inspiring, mentoring, and nurturing young minds. And it's about preparing students for a world where AI is an integral part of daily life and work.

As we look to the future, we can envision classrooms where AI seamlessly augments human teaching, where virtual reality transports students to immersive learning environments, and where personalized AI tutors provide round-the-clock support. We can imagine a world where geographical and socioeconomic barriers to quality education are significantly reduced, where lifelong learning is the norm, and where education is more closely aligned with the rapidly evolving needs of the job market.

However, realizing this vision requires careful navigation of the challenges ahead. We must be vigilant in addressing issues of data privacy and security, ensuring that the vast amounts of data collected by AI educational systems are protected and used ethically. We need to work tirelessly to bridge the digital divide, ensuring that the benefits of AI in education are accessible to all students, regardless of their background or location. We must also be proactive in addressing potential biases in AI systems, striving to create tools that are fair and inclusive for all learners.

The ethical implications of AI in education will require ongoing attention and dialogue. As AI systems take on more significant roles in educational decision-making, from personalized learning paths to college admissions, we must ensure that these systems are transparent, accountable, and aligned with our educational values and ethics. This will require collaboration between educators, ethicists, technologists, and policymakers to develop robust frameworks for the ethical use of AI in education.

Moreover, as AI capabilities continue to advance, we must continually reassess and redefine the goals of education. While AI can excel at imparting knowledge and skills, uniquely human qualities like creativity, empathy, and ethical reasoning will become increasingly valuable. Our educational systems must evolve to place greater emphasis on these human strengths, preparing students not to compete with AI, but to leverage and complement it.

The future of learning with AI also holds exciting possibilities for research and innovation. As AI systems gather and analyze vast amounts of data on how students learn, we have the potential to gain unprecedented insights into the learning process itself. This could lead to breakthroughs in our understanding of cognition, learning disabilities, and effective teaching methods, potentially revolutionizing educational psychology and pedagogy.

Furthermore, the integration of AI in education opens up new avenues for interdisciplinary learning. As AI increasingly crosses traditional subject boundaries, we may see a shift away from rigid subject divisions toward more holistic, project-based learning that better reflects the interconnected nature of real-world challenges. This could foster a new generation of versatile, adaptable thinkers prepared to tackle complex global issues.

The role of teachers in this AI-enhanced future cannot be overstated. While AI will automate many tasks, the importance of human teachers as mentors, facilitators, and role models will only grow. The empathy, inspiration, and human connection that great teachers provide cannot be replicated by AI. Instead, AI will empower teachers to be more effective, allowing them to focus on higher-order teaching tasks and providing them with rich data and insights to inform their teaching strategies.

As we embrace this AI-powered educational frontier, it's crucial to maintain a balance between technological innovation and humanistic values. Education is not just about acquiring knowledge and skills; it's about personal growth, social development, and the cultivation of wisdom. As we integrate AI into our educational systems, we must ensure that these fundamental aspects of education are not just preserved, but enhanced.

The future of learning with AI is filled with boundless possibilities. It offers the potential to create an educational system that is more personalized, more accessible, more effective, and more equitable than ever before. But realizing this potential will require concerted effort, careful planning, and a commitment to ethical, student-centered implementation.

As we stand at this pivotal moment in educational history, we have the opportunity – and the responsibility – to shape an AI-enhanced educational future that empowers all learners, respects human values, and prepares students not just to adapt to the AI-driven world, but to actively shape it for the better. By embracing this challenge with wisdom, creativity, and a steadfast commitment to ethical principles, we can usher in a new golden age of learning, where the power of AI amplifies the best of human teaching and learning.

The journey ahead is complex and challenging, but also incredibly exciting. It calls for collaboration across disciplines, ongoing dialogue between all stakeholders, and a willingness to reimagine education from the ground up. As we embark on this journey, let us be guided by a vision of education that harnesses the power of AI to unlock the full potential of every learner, fostering a generation of knowledgeable, skilled, creative, and ethically grounded individuals ready to thrive in and shape the AI-driven world of tomorrow.

In conclusion, the future of learning with AI is not a distant concept – it's unfolding right now, in classrooms, homes, and online learning environments around the world. Every educator, parent, policymaker, and student has a role to play in shaping this future. By staying informed, engaging in ongoing dialogue, and approaching the integration of AI in education with both enthusiasm and critical thinking, we can all contribute to creating an educational ecosystem that leverages the best of what AI and human intelligence have to offer.

As we look ahead, let us approach the future of learning with AI not with fear or blind optimism, but with informed hope and a commitment to responsible innovation. The AI-powered educational frontier before us holds the potential to democratize knowledge, personalize learning at an unprecedented scale, and cultivate the skills and mindsets needed for success in the 21st century and beyond. It's up to all of us to ensure that this potential is realized in a way that benefits all learners and contributes to a more knowledgeable, skilled, and ethically grounded society.

The future of learning is here, and it's powered by AI. Let's embrace it, shape it, and ensure that it serves the highest aspirations of education – to enlighten minds, empower individuals, and create a better world for all. As we continue to explore and refine the role of AI in education, we're not just changing how we learn – we're reimagining what it means to be educated in the age of artificial intelligence. The adventure is just beginning, and the possibilities are limitless. Let's step boldly into this new frontier, guided by wisdom, empathy, and a shared commitment to unlocking the full potential of every learner in the AI-enhanced educational landscape of tomorrow.

CHAPTER 9

Wrapping Up Your AI Journey

As the sun set on Westfield High School, casting a warm glow through the computer lab windows, Ms. Chen gathered her students for one final session. The holographic display flickered to life, showcasing a montage of their AI-assisted learning adventures over the past year.

"We've come a long way," Ms. Chen began, her voice filled with pride. "From our first tentative steps into the world of AI to the innovative projects you've created, you've not only learned about AI but have become active shapers of its future in education."

The students exchanged excited glances, each recalling their personal triumphs and challenges along this journey. As they prepared to reflect on their experiences and look toward the future, there was a palpable sense of accomplishment and anticipation in the air.

This scene encapsulates the transformative journey that AI has brought to education. As we conclude our exploration of AI in learning, it's crucial to reflect on the key insights we've gained, celebrate the milestones achieved, and chart a course for the continuing evolution of AI-assisted education.

9.1 Key Takeaways: A Comprehensive Review of Our AI-Assisted Learning Odyssey

Throughout this book, we've delved deep into the multifaceted world of AI in education, uncovering its potential to revolutionize learning experiences and outcomes. Let's revisit the critical insights we've gained:

The Power of Prompts: We've discovered that the art and science of crafting effective prompts is fundamental to harnessing AI's potential in education. Prompts are not merely questions or instructions; they are the key to unlocking personalized, engaging, and effective learning experiences. We've learned that well-crafted prompts can guide AI to generate content that is not only informative but also tailored to specific learning objectives, cognitive levels, and individual student needs. The ability to create nuanced prompts that elicit precise and relevant responses from AI systems is a skill that will continue to be valuable as AI technology evolves.

AI Tools and Platforms: Our journey has taken us through the landscape of AI tools like ChatGPT, Claude, and Google's Gemini, revealing their unique capabilities and potential applications in education. We've seen how these tools can serve as tireless tutors, creative muses, and analytical assistants, supporting students and teachers alike in various educational tasks. From generating practice problems and providing instant feedback to facilitating language learning and fostering creative writing, these AI platforms have demonstrated their versatility and power to enhance the learning process.

Ethical Considerations and Responsible AI Integration: A critical aspect of our exploration has been the emphasis on the ethical use of AI in education. We've grappled with important questions about data privacy, algorithmic bias, and the potential impact of AI on equity in education. We've learned that responsible AI integration requires ongoing vigilance, transparent policies, and a commitment to ensuring that AI tools are used to level the playing field rather than exacerbate existing inequalities. The importance of maintaining human oversight, preserving the teacher–student relationship, and fostering critical thinking skills in an AI-enhanced educational environment has been a recurring theme.

The Future of AI-Powered Classrooms: Our journey has allowed us to peek into the future of education, where AI is seamlessly integrated into every aspect of learning. We've envisioned classrooms where personalized learning is the norm, where immersive AI-powered simulations bring abstract concepts to life, and where adaptive assessments provide real-time, actionable insights to both students and teachers. This future holds the promise of more engaging, effective, and inclusive education, but it also challenges us to redefine the roles of educators and the very nature of learning itself.

The Role of Stakeholders: Throughout our exploration, we've emphasized the crucial role that parents, educators, policymakers, and students themselves play in shaping the future of AI in education. We've learned that successful integration of AI in learning requires a collaborative effort, with each stakeholder bringing unique perspectives and responsibilities. Parents and guardians are key to supporting and guiding students in their AI-assisted learning journeys, while educators must continually adapt their teaching strategies to leverage AI effectively. Policymakers have the important task of creating frameworks that promote innovation while safeguarding student interests, and students themselves must develop the skills to use AI tools critically and creatively.

Lifelong Learning and Adaptability: Perhaps one of the most significant takeaways from our journey is the recognition that AI is ushering in an era where lifelong learning is not just beneficial but essential. We've seen how AI is reshaping the job market and creating new opportunities, making it crucial for learners of all ages to continuously update their skills and knowledge. The ability to adapt to new technologies, to learn how to learn with AI assistance, and to maintain a growth mindset in the face of rapid change are skills that will be invaluable in the AI-driven future.

Interdisciplinary Approach: Our exploration has revealed that AI in education is not confined to any single subject or discipline. Instead, it encourages an interdisciplinary approach to learning, breaking down traditional subject boundaries and fostering connections between different areas of knowledge. We've seen how AI can help students see the relationships between science, art, mathematics, and humanities, preparing them for a world where complex problems require multifaceted solutions.

Data-Driven Insights: We've learned about the power of AI to provide unprecedented insights into the learning process. By analyzing vast amounts of data on how students learn, AI systems can offer valuable information to educators, helping them tailor their teaching methods and identify areas where students need additional support. This data-driven approach to education holds the potential to make learning more efficient and effective, but it also raises important questions about data privacy and the balance between personalization and standardization in education.

Creativity and Innovation: Contrary to fears that AI might stifle human creativity, our journey has shown how AI can be a powerful tool for fostering innovation and creative thinking. We've explored how AI can serve as a collaborative partner in creative processes, generating ideas, providing inspiration, and even co-creating with students in areas like art, music, and writing. This collaboration between human creativity and AI capabilities opens up new possibilities for expression and innovation in education.

Global Perspective: Our exploration of AI in education has also highlighted its potential to break down geographical barriers and provide access to quality education on a global scale. We've seen how AI-powered platforms can connect students with resources and experts from around the world, fostering cross-cultural understanding and preparing students for a globally connected future.

9.2 Celebrating Milestones: Recognizing Progress in AI-Assisted Learning

As we reflect on the journey through AI-assisted learning, it's important to recognize and celebrate the significant milestones that have been achieved. These accomplishments not only mark personal growth but also contribute to the broader advancement of AI in education.

Mastery of Prompt Engineering: One of the most tangible skills developed throughout this journey is the ability to craft effective prompts for AI systems. This skill goes beyond simply asking questions; it involves understanding how to structure queries to elicit the most useful and relevant responses from AI. Students and educators who have honed this skill can now engage with AI tools in more sophisticated ways, using them to explore complex topics, generate creative ideas, and solve challenging problems across various subjects.

Ethical Awareness and Responsible Use: A crucial milestone is the development of a strong ethical framework for using AI in education. This includes understanding the potential biases in AI systems, respecting data privacy, and considering the broader societal implications of AI use. Students and educators who have reached this milestone can critically evaluate AI tools and their outputs, ensuring that these technologies are used in ways that are fair, inclusive, and beneficial to all learners.

Integration of AI in Learning Strategies: Another significant achievement is the successful integration of AI tools into personal and classroom learning strategies. This might involve using AI for research assistance, leveraging AI-powered tutoring systems for personalized practice, or employing AI tools for content creation and analysis. The ability to seamlessly incorporate AI into the learning process, complementing traditional methods, marks a major step forward in embracing the potential of these technologies.

Collaborative AI Projects: Many learners have reached the milestone of successfully collaborating on AI-driven projects. This could include developing AI applications, conducting AI-assisted research, or creating AI-enhanced educational resources. Such projects not only demonstrate technical skills but also showcase the ability to work at the intersection of human creativity and AI capabilities.

AI Literacy and Advocacy: An important milestone is the development of AI literacy – the ability to understand, use, and critically evaluate AI technologies. This extends to becoming advocates for responsible AI use in education, participating in discussions about AI policies in schools, and educating peers and community members about the potential and pitfalls of AI in learning.

Personalized Learning Achievements: For many students, a significant milestone is achieving personal learning goals with the assistance of AI. This could be mastering a challenging subject, improving language skills, or exploring advanced topics beyond the standard curriculum. AI's ability to provide personalized learning experiences has enabled many students to progress at their own pace and delve into areas of personal interest.

Cross-Disciplinary Applications: A notable achievement is the ability to apply AI across different subjects and disciplines. Students who can use AI tools to make connections between seemingly disparate fields – for example, using AI to analyze literary texts through a data science lens or applying AI in art projects – demonstrate a sophisticated understanding of AI's versatility.

Innovative Assessment Techniques: Educators who have developed new ways of assessing student learning that incorporate AI have reached an important milestone. This might include using AI to create adaptive assessments, employing AI analytics to gain insights into student progress, or developing AI-assisted portfolios that showcase student work and growth over time.

Global Collaboration: Another significant achievement is participating in global, AI-facilitated learning experiences. This could involve collaborating with students from different countries on AI projects, engaging in AI-powered language exchanges, or contributing to international AI education initiatives.

Mindset Shift: Perhaps the most profound milestone is the development of a growth mindset toward AI and technology in general. This involves embracing lifelong learning, being open to continuous adaptation, and viewing AI as a tool for augmenting human capabilities rather than replacing them.

9.3 Charting the Course: Next Steps in Your AI-Assisted Learning Journey

As we conclude our exploration of AI in education, it's clear that this is not an endpoint but rather a launching pad for future growth and discovery. The rapidly evolving nature of AI technology means that the learning journey is ongoing. Here are some key steps to consider as you continue to navigate the AI-enhanced educational landscape:

Stay Informed and Adaptable: The field of AI is constantly evolving, with new tools and applications emerging regularly. Make a commitment to stay informed about the latest developments in AI and education. This might involve following AI education blogs, attending webinars or conferences, or participating in online courses focused on AI in learning. Remember, adaptability is key in this fast-paced field.

Deepen Your Expertise: Consider specializing in specific areas of AI in education that align with your interests or needs. This could involve diving deeper into natural language processing for language learning, exploring AI in STEM education, or focusing on AI-powered accessibility tools for inclusive learning. Developing expertise in niche areas can provide valuable insights and opportunities.

Engage in Continuous Experimentation: Don't be afraid to experiment with new AI tools and approaches in your learning or teaching practice. Set up small pilot projects to test new ideas, and be willing to learn from both successes and failures. This hands-on approach will help you gain practical experience and develop a nuanced understanding of AI's potential and limitations in education.

Foster AI Literacy: Whether you're an educator, student, or parent, work on developing comprehensive AI literacy. This goes beyond just knowing how to use AI tools – it involves understanding the underlying principles of AI, being able to critically evaluate AI outputs, and considering the ethical implications of AI use. Share this knowledge with others to promote wider AI literacy in your educational community.

Collaborate and Network: Seek out opportunities to collaborate with others who are interested in AI in education. This could involve joining online communities, participating in hackathons or AI education challenges, or forming local study groups. Collaboration can lead to innovative ideas and provide support as you navigate the complexities of AI in learning.

Advocate for Responsible AI Use: Take an active role in shaping AI policies and practices in your educational institution or community. Advocate for ethical AI use, data privacy protection, and equitable access to AI-enhanced learning opportunities. Your voice and experiences are valuable in ensuring that AI is integrated into education in ways that benefit all learners.

Explore Interdisciplinary Applications: Look for ways to apply AI across different subjects and disciplines. This interdisciplinary approach can lead to innovative learning experiences and help prepare students for a future where complex problems require diverse knowledge and skills.

Develop AI-Enhanced Learning Materials: Consider creating your own AI-enhanced learning resources. This could involve developing AI-powered educational apps, creating AI-assisted lesson plans, or designing AI-driven assessment tools. By actively contributing to the field, you can help shape the future of AI in education.

Engage in Action Research: If you're an educator, consider conducting action research in your classroom to study the impact of AI on learning outcomes. Document your experiences, collect data, and share your findings with the broader education community. This kind of practical research is invaluable in understanding the real-world applications and effects of AI in education.

Prepare for Emerging Technologies: While focusing on current AI technologies, also keep an eye on emerging technologies that may intersect with AI in education. This could include developments in virtual and augmented reality, blockchain in education, or advances in brain–computer interfaces. Understanding these emerging fields can help you anticipate future trends in AI-enhanced learning.

Cultivate Soft Skills: As AI takes over more routine tasks, focus on developing and nurturing uniquely human skills. These include emotional intelligence, creative problem-solving, ethical reasoning, and effective communication. These skills will become increasingly valuable in an AI-enhanced world.

Contribute to the AI Education Community: Consider sharing your experiences and insights through blog posts, conference presentations, or academic publications. Your unique perspective can contribute to the growing body of knowledge about AI in education and help others on their learning journeys.

9.4 A Vision for the Future: AI-Enhanced Education for All

As we look to the future, it's clear that AI will play an increasingly significant role in education. However, the ultimate goal should be to harness AI's potential to create more equitable, engaging, and effective learning experiences for all students. Here's a vision of what this AI-enhanced educational future might look like:

Personalized Learning at Scale: AI will enable truly personalized learning experiences for every student, adapting content, pace, and teaching methods to individual needs and learning styles. This personalization will be dynamic, continuously adjusting based on student performance and engagement.

Empowered Educators: Teachers will be augmented by AI, not replaced. AI will handle routine tasks, provide real-time insights into student progress, and suggest personalized interventions, allowing teachers to focus on higher-order teaching tasks, mentorship, and fostering social–emotional skills.

Inclusive and Accessible Education: AI-powered tools will break down barriers to learning, providing real-time translations, adaptive interfaces for students with disabilities, and bridging gaps in educational resources across different regions and socioeconomic backgrounds.

Lifelong Learning Ecosystems: AI will facilitate seamless lifelong learning, connecting formal education with on-the-job training and personal development. Adaptive AI systems will help individuals identify skill gaps and provide tailored learning pathways throughout their careers.

Ethical AI Integration: The use of AI in education will be guided by robust ethical frameworks, ensuring data privacy, algorithmic fairness, and transparency. Students will be educated about AI ethics, preparing them to be responsible creators and users of AI technology.

Global Collaborative Learning: AI will enable meaningful cross-cultural educational experiences, connecting students from around the world for collaborative projects and fostering global citizenship.

Interdisciplinary, Project-Based Learning: AI will support complex, interdisciplinary projects that mirror real-world challenges, helping students develop critical thinking, creativity, and problem-solving skills across multiple domains.

Continuous Assessment and Feedback: Traditional high-stakes testing will be replaced by continuous, low-stakes assessment powered by AI, providing ongoing feedback to students and teachers and reducing test anxiety while improving learning outcomes.

AI-Human Collaborative Creativity: AI will serve as a creative partner in art, music, writing, and design, pushing the boundaries of human creativity and opening new avenues for expression and innovation in education.

Data-Informed Educational Policy: AI-driven analytics will provide policymakers with comprehensive, real-time data on educational outcomes, enabling more responsive and effective educational policies.

Adaptive Physical and Virtual Learning Spaces: AI will optimize both physical classrooms and virtual learning environments, adjusting lighting, temperature, and layout in physical spaces, and personalizing virtual environments to maximize student engagement and learning.

Predictive Support Systems: AI will help identify students at risk of falling behind or dropping out, enabling early interventions and support to ensure every student has the opportunity to succeed.

This vision of AI-enhanced education is not without challenges. It will require ongoing collaboration between educators, technologists, policymakers, and ethicists to ensure that AI is integrated into education in ways that are equitable, ethical, and truly beneficial to all learners. It will also require a commitment to ongoing research, experimentation, and adaptation as we learn more about the impacts of AI on learning and development.

As we conclude this journey through the world of AI in education, it's clear that we are standing at the threshold of a new era in learning. The potential of AI to transform education is immense, but so too is our responsibility to shape this transformation in ways that align with our highest educational ideals and values.

The future of AI in education is not predetermined; it will be shaped by the choices we make today and in the years to come. By staying informed, engaged, and committed to ethical and equitable AI integration, we can work toward a future where AI enhances human potential, fosters lifelong learning, and helps create a more knowledgeable, skilled, and empathetic global society.

As you continue your journey with AI in education, remember that you are not just a passive recipient of this technology, but an active participant in shaping its future. Your experiences, insights, and advocacy will play a crucial role in ensuring that AI serves as a powerful tool for educational empowerment and social progress.

Thank you for joining this exploration of AI in education. May your ongoing journey be filled with discovery, innovation, and meaningful learning experiences. The future of education is in our hands, and with AI as our ally, the possibilities are boundless.

Exploring Generative AI for Art, Graphics, and Videos in Education

A.1 Introduction

As we venture deeper into the world of AI in education, it's crucial to explore the creative potential of generative AI tools. This appendix is dedicated to introducing educators, parents, and students from 3rd grade to 12th grade to the exciting possibilities of using AI to create art, graphics, and videos. We'll explore practical applications of tools like Midjourney, DALL-E, Gemini, and Claude, providing step-by-step examples and exercises tailored to different age groups and educational contexts.

P. Narciso, *Generative AI in Education*, https://doi.org/10.1007/979-8-8688-0844-9

A.2 Understanding Generative AI for Visual and Video Creation

Generative AI for visual and video creation uses machine learning algorithms to produce new images, animations, and videos based on text prompts or existing visual data. These tools have opened up new avenues for creativity and expression, allowing users to bring their imaginations to life in ways previously unimaginable.

Key Tools:

1. Midjourney: A text-to-image AI tool known for its artistic and often surreal creations.

2. DALL-E: OpenAI's image generation model, capable of creating a wide range of images from textual descriptions.

3. Google's Gemini: While primarily a text-based AI, Gemini has multimodal capabilities, including understanding and describing images.

4. Anthropic's Claude: Similar to Gemini, Claude can analyze and describe images but cannot generate them.

Note   As AI technology rapidly evolves, the capabilities of these tools may change. Always refer to the most current information about each tool's features and limitations.

Examples and Exercises by Grade Level

3rd–5th Grade: Introduction to AI Art

For younger students, the focus should be on sparking creativity and understanding the basic concept of AI-generated art.

A.3 Accessing AI Tools: Step-by-Step Instructions

Before we dive into the exercises, let's go through the process of accessing each AI tool we'll be using.

Accessing DALL-E:

1. Go to OpenAI's website (`https://openai.com/dall-e-2`).

2. Sign up for an account or log in if you already have one.

3. Navigate to the DALL-E interface.

4. You'll see a text box where you can enter your prompt.

Accessing Midjourney:

1. Join the Midjourney Discord server (`https://discord.gg/midjourney`).

2. Go to one of the #newbies channels or start a new thread in #general.

3. Type "/imagine" followed by your prompt to start generating images.

Accessing Google's Gemini:

1. Go to the Gemini AI website (`https://gemini.google.com/`).

2. Sign in with your Google account.

3. You'll see an interface where you can enter text and upload images for analysis.

Accessing Anthropic's Claude:

1. Go to the Claude AI website (`https://www.anthropic.com/` or the specific Claude interface provided by Anthropic).

2. Sign up for an account or log in.

3. You'll see an interface where you can enter text and upload images for analysis.

A.4 Examples and Exercises by Grade Level

A.4.1 3rd–5th Grade: Introduction to AI Art

Exercise 1: Imaginative Creatures with DALL-E

Objective: Introduce students to the concept of AI-generated images by creating fantastical creatures.

Steps:

1. Access DALL-E using the instructions provided above.

2. Have students describe an imaginary creature, combining features of different animals.

3. Help them formulate a clear, descriptive prompt. For example: "A friendly creature with the body of a lion, wings of an eagle, and a unicorn's horn, playing in a colorful meadow."

4. Enter this prompt into DALL-E's text box.

5. Click the "Generate" button.

6. DALL-E will produce several image options. Have students select their favorite.

7. Discuss how the AI interpreted their descriptions and what surprised them about the image.

Exercise 2: Illustrating Stories with DALL-E

Objective: Use AI-generated art to illustrate short stories or poems written by students.

Steps:

1. Have students write a short story or poem (4–6 lines).

2. Access DALL-E.

3. Help them create prompts for key scenes in their story. For example: "A magical treehouse floating in a starry night sky, with a rainbow bridge leading to it."

4. Enter this prompt into DALL-E.

5. Click "Generate" to create the image.

6. Repeat for other key scenes in the story.

7. Use a simple design tool or even a word processor to combine the students' text with the AI-generated illustrations.

8. Create a digital or physical book combining the students' text and the AI-generated illustrations.

A.5 6th–8th Grade: Exploring AI Art Techniques

Exercise 3: Historical Figure Portraits with DALL-E

Objective: Create portraits of historical figures in various artistic styles to enhance history lessons.

Steps:

1. Choose a historical figure being studied in class.

2. Research the figure's appearance and the artistic styles of their era.

3. Access DALL-E.

4. Craft a detailed prompt combining the figure and style. For example: "A portrait of Leonardo da Vinci in the style of a Renaissance oil painting, with symbols of his inventions in the background."

5. Enter this prompt into DALL-E.

6. Click "Generate" to create the image.

7. If the result isn't quite right, try refining the prompt. For example: "A highly detailed portrait of Leonardo da Vinci with long hair and beard, wearing Renaissance-era clothing, surrounded by sketches of his inventions, in the style of a 15th-century Italian oil painting."

8. Generate again with the refined prompt.

9. Compare the AI-generated portrait with actual historical portraits.

Exercise 4: Science Concept Visualization with DALL-E
Objective: Use AI to create visual representations of scientific concepts.
Steps:

1. Select a scientific concept being studied (e.g., photosynthesis, the water cycle, or plate tectonics).

2. Access DALL-E.

3. Develop a prompt that describes the concept visually. For example: "A detailed cross-section of a leaf showing the process of photosynthesis, with sunlight, chlorophyll, and glucose production visible."

4. Enter this prompt into DALL-E.

5. Click "Generate" to create the image.

6. If needed, refine the prompt for more accuracy. For example: "A scientifically accurate diagram of a leaf cross-section during photosynthesis, showing sunlight entering, chloroplasts absorbing light, and the production of glucose, with labels for each part of the process."

7. Generate again with the refined prompt.

8. Have students analyze the image, identifying accurate representations and any misconceptions.

A.5.1 9th–12th Grade: Advanced AI Art and Video Projects

Exercise 5: Social Commentary Art with DALL-E

Objective: Create thought-provoking images addressing current social issues.

Steps:

1. Have students research a current social or environmental issue.

2. Access DALL-E.

3.   Develop a concept for an artwork that comments on this issue.

4.   Craft a detailed prompt that conveys the concept. For example: "A powerful image showing the impact of plastic pollution on marine life, with a sea turtle swimming through an ocean filled with colorful plastic waste, in the style of a surrealist painting."

5.   Enter this prompt into DALL-E.

6.   Click "Generate" to create the image.

7.   If the result doesn't fully capture the concept, refine the prompt. For example: "A hyper-realistic image of a sea turtle entangled in plastic waste, swimming through a once-beautiful coral reef now bleached and covered in discarded plastic bottles and bags, with a contrasting split-screen showing a healthy reef, symbolizing environmental degradation."

8.   Generate several variations using the refined prompt.

9.   Have students select the most impactful image and write an artist's statement explaining their concept and how the AI-generated image conveys their message.

Exercise 6: AI-Assisted Video Creation with DALL-E Images
Objective: Combine AI-generated images with video editing to create a short film.
Steps:

1.   Develop a storyboard for a 1–2 minute video on an educational topic.

2.   Access DALL-E.

3. For each scene in your storyboard, create a prompt describing the image needed. For example: "A classroom of the future with holographic displays, robots assisting students, and children of diverse backgrounds collaborating on a science project."

4. Generate images for each scene using DALL-E.

5. Refine prompts and regenerate images as needed to get the desired visuals for each scene.

6. Download all the generated images.

7. Use video editing software to combine the AI-generated images with transitions, text overlays, and audio narration.

8. Add background music and sound effects as needed.

9. Export the final video and present it to the class.

A.6 Cross-Grade Level: Analyzing Images with Gemini and Claude

Exercise 7: AI Art Analysis

Objective: Use AI to gain insights into artworks and develop critical analysis skills.

Steps for Gemini:

1. Select an artwork appropriate for the grade level.

2. Access Gemini using the instructions provided earlier.

3. Upload the image to Gemini.

4. Ask Gemini to describe the image in detail. Type: "Describe this image in detail, focusing on the composition, use of color, and overall mood."

5. Based on Gemini's response, ask follow-up questions. For example: "What artistic style does this image appear to be in? What are the characteristic elements of this style?"

6. Continue the conversation, asking about interpretation, historical context, or comparisons to other artworks.

Steps for Claude:

1. Select the same artwork used with Gemini.

2. Access Claude using the instructions provided earlier.

3. Upload the image to Claude.

4. Ask Claude to analyze the image. Type: "Please analyze this image. Describe what you see, including the subject matter, style, colors, and any notable features."

5. Based on Claude's initial response, ask for deeper analysis. For example: "Based on what you've described, what emotions or themes do you think this image is trying to convey? How does the artist use color and composition to achieve this?"

6. Compare Claude's analysis with Gemini's and discuss the similarities and differences.

A.7 Ethical Considerations and Best Practices

As we explore the use of AI in art and video creation within educational settings, it's crucial to address ethical considerations and establish best practices:

1. Attribution and Copyright: Discuss the importance of properly attributing AI-generated content and understanding the copyright implications of using AI-created art in various contexts.

2. Bias and Representation: Examine how AI art generators might perpetuate or challenge societal biases in their representations of people, cultures, and concepts.

3. Balancing AI and Human Creativity: Emphasize that AI tools should enhance, not replace, human creativity. Encourage students to use AI-generated content as inspiration or components of larger, original works.

4. Digital Literacy: Teach students to critically evaluate AI-generated content, understanding its capabilities and limitations.

5. Ethical Prompt Crafting: Guide students in creating prompts that are respectful, inclusive, and do not promote harmful stereotypes or inappropriate content.

6. Data Privacy: Discuss the importance of data privacy when using online AI tools, especially when uploading personal images or information.

A.8 Conclusion

The integration of generative AI tools for art, graphics, and video creation in education offers exciting opportunities for enhancing creativity, visual literacy, and cross-curricular learning. By engaging with these tools thoughtfully and ethically, educators can provide students with valuable skills for navigating and contributing to an increasingly AI-influenced world.

As AI technology continues to evolve, it's important to stay informed about new developments and continuously reassess how these tools can best serve educational goals. The exercises and examples provided here are just a starting point – the true potential of AI in art education will be realized through the innovative approaches of curious and dedicated educators and students.

Remember, the goal is not to replace traditional art education but to expand the creative possibilities available to students. By combining AI tools with traditional techniques, we can foster a new generation of artists, designers, and visual thinkers equipped to express themselves in both digital and physical realms.

Glossary

Adaptive Learning: An educational method that uses AI algorithms to adjust the learning experience based on individual student performance and needs.

AI Literacy: The ability to understand, use, critically evaluate, and engage with artificial intelligence technologies.

AI Model: A computational representation trained on data to recognize patterns and make predictions or decisions.

Algorithmic Bias: Systematic errors in AI systems that can lead to unfair outcomes for certain groups.

Artificial Intelligence (AI): Computer systems designed to perform tasks that typically require human intelligence, such as visual perception, speech recognition, decision-making, and language translation.

Automated Essay Scoring (AES): AI-powered systems that evaluate and provide feedback on written essays.

Chatbot: An AI-powered program designed to simulate human-like conversation through text or voice interactions.

Claude: An AI assistant created by Anthropic, designed for open-ended conversations and various tasks.

Computer-Supported Collaborative Learning: An educational approach that uses technology to facilitate group learning and interaction.

Data Privacy: The protection of personal information collected and used by AI systems and other technologies.

Digital Divide: The gap between those who have access to modern information and communication technology and those who do not.

Ethical AI: The development and use of AI systems in ways that adhere to moral principles and values.

P. Narciso, *Generative AI in Education*, https://doi.org/10.1007/979-8-8688-0844-9

Explainable AI: AI systems that provide clear explanations for their decisions or outputs, making them more transparent and understandable to users.

Formative Assessment: Ongoing evaluation of student learning to provide feedback and inform instruction.

Generative AI: AI systems capable of creating new content, such as text, images, or music, based on patterns learned from existing data.

Gemini: Google's advanced AI model designed for multimodal tasks, capable of understanding and generating text, code, images, and audio.

Human in the Loop: An approach to AI that keeps humans involved in the decision-making process, providing oversight and intervention when necessary.

Immersive Learning: Educational experiences that use technologies like virtual or augmented reality to create highly engaging, interactive environments.

Intelligent Tutoring System (ITS): AI-based educational software designed to provide personalized instruction and feedback to students.

Large Language Model: An AI model trained on vast amounts of text data, capable of understanding and generating human-like text.

Machine Learning: A subset of AI that focuses on the development of algorithms that can learn from and make predictions or decisions based on data.

Natural Language Processing (NLP): A branch of AI that deals with the interaction between computers and human language.

Personalized Learning: An educational approach that tailors instruction to individual student needs, goals, and preferences.

Prompt Engineering: The skill of crafting effective instructions or questions for AI systems to generate desired outputs.

Responsible AI: The practice of developing, deploying, and using AI systems in ways that are ethical, transparent, and accountable.

Universal Design for Learning (UDL): An educational framework aimed at designing learning experiences that are accessible and effective for all students, regardless of their abilities or background.

Virtual Reality (VR): A computer-generated simulation of a three-dimensional environment that can be interacted with in a seemingly real way.

Index

Printed in the United States
by Baker & Taylor Publisher Services